Gabriel Dima

The Evolution of the Aerostructures Concept and Technologies

Editor: Michael M. Dediu

DERC Publishing House

Tewksbury (Boston), Massachusetts, U. S. A.

Published and printed in the
United States of America

Library of Congress Cataloging in Publication Data

The Evolution of the Aerostructures. Concept and Technologies

ISBN-13: 978-1-939757-29-6

Scientific reviewers:
Prof. Dr. Ioan Curtu, Prof. Dr. Ion Balcu, Prof. Dr. Ionel Staretu

Translation: Dr. Eng. Diana Cazangiu, Eng. Mircea Zamfir

Desktop publishing: Roxana Ciobanu

Corrections: Eng. Mircea Sasu, Eng. Paul T. Dinu, Eng. Mihai
Achim

Cover producer: Alexandru Maris

EDITOR'S NOTE

After receiving many favorable comments and having interesting discussions regarding the previous books published by us, we are pleased to present this new book, which will certainly be a delight for everyone interested in aviation, from the earliest dreams of flying – Daedalus, in Greek mythology, who constructed wings for his son, Icarus, literarily explained by Publius Ovidius Naso (43 BC – 17 AD) in his Metamorphoses (8 AD) – then the first design of flying machines by Leonardo da Vici (1452 – 1519) – to the present day advanced aircrafts, UAVs and spacecrafts.

We thank Mrs. Sophia Dediu and our team for their assistance in preparing this book.

There is, surely, much more that can be said about this subject, and we hope that this book will provide ideas for our audience, which will stimulate more research, development and applications.

We look forward to receiving comments and suggestions from our readers.

Michael M. Dediu, Ph. D.

Boston, October 4, 2015

Previously published in this series:

1. Ioan Goia, *Mechanics of Materials*
2. Ionel Staretu, *Gripping Systems*
3. Proceedings of the 4th International Conference *"Advanced Composite Materials Engineering"* COMAT 2012,
4. Proceedings of the 5th International Conference *"Computational Mechanics and Virtual Engineering"* COMEC 2013
5. Proceedings of the 5th International Conference *"Advanced Composite Materials Engineering"* COMAT 2014,
6. Sorin Vlase: *Mechanical Identifiability in Automotive Engineering*

Foreword

The aviation, carrying the aspirations of the human spirit, brought unprecedented satisfactions, but in the same time unforgiving with the human error. Built on human sacrifices with a long time before the revolution brought by the first flight, it succeeded, only due to the tenacity of its pioneers, to create with small steps, an industry that, in only 50 years, has reached to complement and to compete with the recognized transport ways.

The aviation history books rarely describe the economic realities of the times. As a business, the aviation has always been in a fragile balance. The very large investments, the materials and the technologies being of the knowledge limit, a future hard to foreseen, all these have generated a climate of risk, which was looked with skepticism by the investors who preferred more "earthly" business. To these is added the business sensitivity having as objective the aerospace production or transport to the flight events. Due to their dramatic nature, these were at the forefront of the mass media, a fact which has propagated to nowadays, having an overwhelming influence over the business. Even in these days we are the witnesses of an airline which is in great difficulty after two tragedies occurred in the period of a few months.

The project management of an aircraft development program is again a very unstable area, due to the enormous budgets, to the long periods (in some cases for more than 10 years), affected by the inherent financial and economic instabilities. In many countries, the companies which have seen a spectacular growth have been bought or nationalized by the State, to be privatized when these were in financial difficulty, etc. On the other hand, the enormous amounts, consistent percentages of GDP of a country, required to equip an air force, led to the finding of interesting solutions such as the programs of manufacturing under license, offset or leasing. When these solutions proved to be

5

exceeded, some countries have teamed up to lead the large scale programs. In the recent years, a new concept is developed, that belongs to the profile clusters, which apparently is the most stable and the most flexible, involving, in addition to the producers, the research institutes, the universities, the services providers and local authorities.

All of these inevitably put their footprint on the development of materials, technologies and the aircraft design. In the aerospace industrial landscape marked by syncopations and many contradictions, it appears a necessary book, about a neglected aspect by the aviation history books. Being rather in the shadow of the aerodynamics, the engines, the weapons and even in-flight operations, the aerospace structure has occasionally received small notes or short comments.

Treating a strictly technical field, for the first time, the author has succeeded, using a language adapted to the general public, to bring to light at such a scale and details, the time evolution of the aircraft structure.

The paper is structured in five parts. The first part familiarizes the reader with the terminology, the basic components, the materials and the particularities of an aircraft. The second part is dedicated to the aerospace structures evolution. The approach is made on identified periods by the priorities, trends and the aircraft characteristics. Finally there are reviewed the present research concerns and the main challenges of the future aircraft. In the third part, there are presented the main structural concepts for wing and fuselage, with the advantages and disadvantages and the key points of their evolution. The evolution of the aircraft design and calculation are investigated in the fourth part, along with the development of the tools used by the development teams. There are analyzed the accidents which have generated the most important changes in the structural health concepts and the requests of the certification process of the civil and military aircrafts, and the interference with other sciences and industries. The last chapter is dedicated to the structural details, namely the latticed beam joints and the monocoque respectively.

The paper presents special topics such as the transition from canvas and wood to aluminum and composite structures, or the historical dispute between the monocoque and the latticed beam. There are presented the main landmarks of the aircraft structures evolution and how the materials, the manufacturing technology, the design, calculation and structural security concepts were developed, especially those related to the damage tolerance and fatigue. There are also dealt some highly topical issues such as the trends in lightweight design (design for minimum weight) or the topological optimization, which already concerns for years the automotive industry.

The paper is useful for students, technical and engineering staff, but in the same time for all who are passionate about technique history and especially of aviation. For the success of this project, the author has made a huge documentation work, data analysis and assessment studies, together with the synthesis, interpretation and valuable conclusions. The author, with experience in the aviation field, proved, in addition to an authentic scientific approach, a technique to communicate easily with the non- specialized readers, but also the art of making an enjoyable and interesting reading.

I consider that the original work and a high held engineering will succeed to reach the goal, capturing the interest of a broad spectrum of readers. This is an integral part of a long term activity of the author in training students and young engineers.

Werner Braun

President of The Transylvania Aerospace Cluster

This book has informative historical knowledge about aircraft structure progress from the very earliest flying machine. From this book we learn in the very early days how the aviators made every possible effort to make aircraft structures strong enough as well as light-weight. All these efforts are not to be taken for granted: in the past all the aircraft structural engineers struggled to make airframe strong beginning with wood materials. Today, we all use high strength aluminum materials for primary airframes to save structural weight. In addition, strength and light-weight is not enough; we have to consider safety, including design considerations of fatigue and fail-safe to safeguard the structural integrity. All these considerations came from extensive testing results and the modified stress analysis (or strength analyses) equations that we use today. These are the so-called "empirical equations" which are very important and are widely used in our aircraft stress analysis today.

I strongly recommend aviators to read this book to understand the aircraft structural progress, which will help all the aviators to understand the importance of the major rule of structures. My airframe books cover up-to-date technologies that meet all the structural requirements in the aircraft arena For the cost-effective aircraft, today's aircraft structural engineers should consider not only strength, weight, safety, cost, material selection (e.g., composite for future light weight), but also maintenability and repairability, as well as the manufacturing cost of the cost-effective aircraft.

Prof. Michael C.Y. Niu

Author's preface

The responsibility is transmitted to the present day, inclusively, within the internships, the students being learned to cultivate an attitude of maximum importance to ensure the airworthiness and safety of all in-flight operations.

For the engineers and technicians who work in aviation, especially those responsible for the structures design and calculation, the knowledge of the history of the progress and the aerospace structures technologies will make them understand the present in an easier way and to better estimate the directions where the aerospace structures will evolve in the future. They will understand that the manufacturing technologies, the standards, the procedures, the work methodologies represent an extremely valuable capital, the experience of several generations of specialists who have sought to leave in an accurate form to the future generations the learnt lessons and the work's best practices. Professor Michael C.Y. Niu, the holder of the most known and used books in the design teams, wrote, in the late 1990s, that young engineers are spending more and more time in front of the computers, neglecting the teamwork with the more experienced "old timers". At the same time, he regrets that the top management of the companies is increasingly concerned about costs reduction, rather than paying attention to the collection and the dissemination of know-how in the form of design manuals and work methodologies.

I consider this work useful even for those who run businesses or manage aviation clusters, but also the governmental authorities in order to be able to elaborate strategies for long-term development based on the valuable lessons of the past.

The future generations should be educated in a spirit of respect for the predecessors, especially for the roads openers. The pioneering period, marked by failures and sacrifices, had the most important role in the aviation development because it was the one which proved that human can overcome its condition and he can follow his dream for thousands of years – the flight.

In any field, the progress is assured not only by the winners - the great discoverers or inventors. Those who research the directions which lead to failures, in fact, they discover new limits. If the success has an unknown direction, the failures are those that reduce the number of unknowns, consisting in certainties, the boundaries of the path on which it needs to go forward.

Unfortunately, the many people who investigate, who invested the energy and the resources of a lifetime, remain in the shadows, the history mentioning only by those who have chosen the right direction. Others remain forgotten, in bankruptcy, being mocked by the journalism of that time. An example is one of the "Bloopers" like broadcast where the viewer was invited, on an "adequate" musical background to laugh of the accidents of those venturesome.

By this book, I wish to pay tribute to all those who, through their passion and hard work of a life, we owe the facilities, opportunities and, why not, the pleasure of flight. I hope that this book will be a modest contribution in informing the young people who work in aviation, but also for those passionate about.

I had the opportunity to work in aviation, to concept, to calculate, to manage the manufacturing and to test the aircrafts components. But more important than this, was the opportunity to meet the people from this area - professors, designers, tooling and manufacturing engineers, quality supervisors, managers, and, last but not least, pilots from which I learned this noble profession. I want to thank, by this way, to the regretted prof. Alexandru Todicescu, prof. Adrian Postelnicu and eng. Şerban Seculin. Also I want to thank to prof. Ioan Goia, prof. Ioan Curtu, prof. Gheorghe Deliu, eng. Gheorghe Ilie, eng. Viorel Ţigău.

I want to bring special thanks to prof. Ion Balcu who had the idea of this book and encouraged and inspired me to finish this work. Many thanks to eng. Radu Ionescu, who relentlessly continues, with the energy and the fervor of a young spirit, an activity for over 40 years dedicated to the aviation.

Finally, I would like to add, that, the work on this book was an extraordinary experience, an opportunity to find out more about the brilliant aviation people and to learn from the lessons of the past, which gave me countless occasions of meditation.

February 2015,   Gabriel D. Dima

About the author

Aviation engineer since 1996, he was formed as a designer at IAR Braşov, where he worked as a project engineer and project manager for different programs for the conversion and the modernization of Puma helicopters and flight simulators for Mi 17 and Mi-24. He continued as design team leader for the Augusta helicopters, designer and checker for Gulfstream and Airbus. Since 2010, he is the co-founder of Nuarb Aerospace Company, where he worked as a stress engineer and technical manager, for main customers from EU, the Russian Federation and China.

Also, he was involved in different research projects, and consultant for companies from the automotive industry in the area of the project management, lightweight design, kinematic analysis and optimization of the manufacturing costs. He completed his PhD stage between 2012 and 2015 in the area of the topology of the lightweight welded structures. He is the author of over 10 inventions, 25 articles and scientific papers in the field of the aerospace structures, lightweight design and topological optimization. He taught subjects as the commercial airplane structure, the project management and different CAD/ CAE software.

CHAPTER 1

Conception and design of the aerostructures

Although, it appeared the last one in comparison with the naval or land transportation, the aviation succeeded in retrieving this handicap of the position of the *new entry*, finally offering a comparable range, speed and ceiling which seemed unthinkable at the beginning of the last century. The growing requests of the civil and military operators leaded to a constant effort to develop more and more efficient aircrafts, which stimulated the development of new design methods and innovative technologies, respectively.

The challenges for the future are considered the limited useful load, the big development and operating costs, and a reduced environmental footprint.

1.1. INTRODUCTION

Starting from a sort of flying machines which were considered by its contemporaries some eccentricities or curiosities, the aviation pioneers proved a special commercial feeling, and in less than 15 years from the first flight, they managed to give airplanes a main role in the First World War. After 1950, the long distance travels or many military missions were unthinkable without aviation, this being considered as a cornerstone for the conquering of the outer space.

Most people admire the unstoppable progress of the aviation in a very short period of time, but in the same time others neglect the period before the first controlled flight since 1903. It can be stated that the people's need for mobility was manifested in all environments; if the seas and the land were early dominated, the air domination needed a longer "incubation" period.

1.1.1 Aviation and the transportation industry

Emerged only 110 years ago, the aviation proved its utility since the first years, starting with the mail and passenger transport and continuing with the freight transport. Being considered the most dangerous way of transport, marked by highly dramatic accidents, the aircraft reached, in a record time period, the position of the safest way of transport (according to the statistics of International Civil Aircraft Organization ICAO). This leading position is the result of a continuous and coordinated effort of learning by mistakes and by the wise and carefully restriction of the certification regulations. Another contributing factor was to search the cause of an accident, no matter the costs and the investigation range; the lessons learned were considered a common good for all the producers from the entire world.

Starting from aircrafts which transported only one passenger, after the WWII, the numbers of the passengers increased significantly, when the long-range bombers were reconverted in passenger aircrafts. Starting on this basement, aircrafts like Boeing 747 and IL 86 with capacities of 550 passengers and 350 passengers were developed only two decades after. In 2005, the Airbus Company rolled out the A380 model, which can take on board up to 850 passengers. It seems that this number, although it was firstly accepted, is not calibrated for the passengers' air transport, as long as A380 didn't record the expected market success. Other reference good to mention is the

Concorde aircraft, which made history by its speed of 2100 *km/h*, remaining the fastest transport airplane till our days [41].

Related to the freight transport, this is on second place, being used for special and mail transports, the other cheaper transport ways being preferred for the consumer goods, raw materials, etc.,.

1.1.2   The role of the aviation in defense

The factors of progress in the aviation were the military and the civil requirements. The demands of the Defense Ministries were the first to transform the aviation from a hobby of some dreamers to a controllable and robust tool, having a defined operability and transport capacity, destined not only to the elite, but also to pilots with a common level of abilities. From another point of view, the commercial aviation considered the aircraft as a business, this being a strong stimulus for the performance improvement by design innovation and refining.

The military aircraft covers a wide range of missions. In the beginning, it appeared as a recognizing mean, but the airplane became in short time a bomber and then a fighter. Nowadays, the airplanes are extremely specialized, reaching up to the level of radar or tanker aircrafts.

The aircraft found its place in the assembly of military forces, alongside the marines and land forces. On the battlefield, the main concern was getting aerial supremacy, which allowed the control of a territory with much lower resources. Starting with the 1950, the helicopters were introduced; these were used in support missions (Vietnam War), following a distinct evolution of the assault and anti-tank aircraft.

In the last period, one focuses on the use of military aviation for humanitarian missions or for intervention in case of disasters (food and equipment transport), search and rescue and medical evacuation.

1.1.3 The utilitarian aviation

Known as the general aviation, the utilitarian aviation includes the private and sports aviation, special missions and emergency situations. The utilitarian aviation executes a wide range of missions for aerial photos and surveillance, external carry transports (containers, equipment, cars, high voltage pillars), the maintenance of the high voltage lines, pesticides application over agricultural crops or the servicing of offshore platforms. The emergency situations demand different missions of search and rescue, medical evacuation, the intervention in the case of fire and the intervention in the situations of the humanitarian crisis. Beside the utilitarian role, the general aviation represents a special niche for the businessmen mobility, making possible to travel on a distance of 13000 *km* in less than a day.

Due to the diversity of missions, scenarios and different weather conditions, the utilitarian aviation requires some dedicated aircrafts with embedded equipment and very well trained staff. Because of the very expensive operating costs, the trend is to design multifunctional aircrafts which may be reconfigured very fast, depending to the specific mission. This capability is common in military aircrafts, which in the case of natural disasters have to be compatible in operating with aircrafts destined for the intervention in emergency situations.

In this area, the helicopters have a special role, with a multitasking profile, allowing the fast reconfiguration of the onboard equipment according to different missions.

As alternative flight principles, the convertible planes like V-22 Osprey (a hybrid between helicopter and the airplane) and the gyrocopters (aircrafts with a rotary wing and propeller) for the transport of $1 \div 4$ persons are used (with limited applicability).

1.2. BRIEF HISTORY OF THE AVIATION

A brief review of the most important events from the aviation history it is made in the present subchapter, taking also into account the precursory period of the first controlled flight from the 1903. The minimum requirements that were necessary for this first flight are highlighted; it is also shown how the inventors from different countries contributed to this flight during the time. The contributions of the Romanian pioneers are highlighted – in the way they are perceived by the international organizations and the main reason why their names are not significant in present the aviation history. Finally, some trends and priorities of the worldwide aeronautics are mentioned.

1.3. THE PIONEERS

The aviation is perhaps the branch of technology which generated the most chronologies and historical papers, many nations disputing their primacy related to the key moments of the air conquest or recognition of the inventions, which transformed different attempts in a flying machine. This public interest was very early, according the statistics of Smithsonian Institute, in 1909, 13500 articles and books on the air machines topic being recorded.

The first successful test of a human flight attested by documents was the one of the balloon of Montgolfier brothers, at Paris, in 1783 November. The balloon hovers by the principle of the Archimedes' force, its use being limited by the volume of gas or warm air which is lighter than the atmospheric air. The balloon hovering doesn't represent a controlled flight, this being directed by the atmospheric currents and wind, the altitude being the single one controllable parameter. The first practical use for the balloons was made in 1794 by the French army, having the role of aerial observation [Antoniu, 2003].

Most of the authors (including [14], [04], [71]) recognize the influence of Sir George Cayley (1773 – 1857) for further development of the aviation. In 1779, he stated that, for taking-off, a flying machine *"heavier than the air"* needs separate devices for lift and propulsion, respectively; till this, the most attempts tried to imitate the birds flight (the wing has both roles of lift member and propulsion). On this line, in 1804, he designed an airship powered by a propeller.

The flying idea was associated for a very long time with the birds flight; the inventors of different flying machines searching a simple imitation, mostly by the manufacture of some wings. Although, many flying configurations or *"techniques"* were recorded during the time, these were not successful because of the biomechanical reasons: the gravity center of the human should be in the right of the sternum, and the chest muscles would have to represent a significant percent by the entire muscles mass. Although, since the first millennium, the humanity had well-defined notions of the structures mechanics, put into practice in constructions, bridges, boats or battle vehicles, regarding the inventions destined to the flight are distinguished by a poor structure, mostly improvised.

In 1842, William Henson constructed a first flying machine (airplane) using the configuration defined by Cayley, which was equipped with a propeller powered by a steam engine of 20 *HP*, this being unable to take off. The aviation history mentions further realization of many flying machines, with no results, outlining the idea that it needed a more appropriate propulsion system and a more efficient wing [41].

One of the most respected researchers of this period was Otto Lilienthal (1848 – 1896). Being an engineer and proving consistency and a good organization, he publishes in 1889 the paper: *"The bird's flight as a base for the flight art"* where he stated that an aircraft heavier than the air can fly. Note the term *the flight art* suggesting the strong belief that the flight is not for everyone.

Between 1890 and 1896, in order to test his ideas and theoretical assumptions, he built 18 gliders used for over 2000 flights, taking-off from an artificial slanted plane. Starting with a wing structure inspired from the bird's wing skeleton, he continued with the radial structures and then with biplanes which were easier controllable, realizing flights of up to 300 meter in distance [41]. Scientifically approaching the flight problem, Lilienthal understood the importance of the controlled flight. He performed experiments with a rotating wing and realized the first tables with the lift and drag values. These tables have been further proved to be inaccurate by repeated experiments performed by the Wright brothers [09].

According to [71], the greatest merit of Otto Lilienthal was to prove that the human can fly. It's good to note, for the Lilienthal's case, the systematic approach of the research, the gliders design, manufacture and flight testing. His dedication for the flight conquest cost him his life; after the fatal accident, its last words were: "*sacrifices must be done*" [04]. Not by chance, in [49], the period before the first flight is called "*the heroic period*".

In 1898, Santos Dumont performed the first flight with an airship, constructing many other models till he turned his attention to the airplanes [41]. In 1900, Count von Zeppelin accomplished the first flight with a large size airship, with rigid structure, inaugurating a period of 40 years of intense exploitation of these flying giants.

In 1897, Clement Adler built a first aircraft (Fig, 1.1, a) based of the product technical and operational requirements realized by the French Ministry of Defense – being able to take onboard 1 – 2 people, material, fuel, ammunition, a flight altitude of few tens of meters and a minimum speed of 55 *km/h*. The airplane, with twin engines, equipped with foldable wings, failed the take-off inside of the official flight due to the strong wind, losing the financial support promised by the French army.

Another famous failure belongs to Samuel Langley – astronomer and physicist, who after some good experiments with some model planes constructed an airplane with tandem wings, with a star engine which powered two propellers (Fig. 1.1, b). Following to be launched from a floating slanted plane, the airplane was hanged from a cable and the take-off failed. Due to this accident, the American Government withdrew its support and Langley had no way for continuing his tests.

The pioneering period has ended in December 1903, when the Wright brothers succeed the first controlled flight, inaugurating a proficient period in the flying machines development.

a)

b)

Fig. 1.1

The airplane of Clement Adler [71]

The Langley airplane model, at 1:4 scale during the flight (1903) [64]

1.3.1 The Romanian contributions

As part of Europe, Romanian inventors showed a particular interest for flight. According to [03], [56] an impressive number of inventors, producers or workshops are mentioned. In the dedicated international literature the following Romanian events are mentioned:

- in 1906, Traian Vuia (France) realized the first take-off by its own machine meanings (the flight wasn't approved); for the first time a landing gear with tires was used (Fig. 1.2);
- in 1907, Paul Cornu realized the first take-off followed by a short vertical flight with a flying machine heavier that the air (the forerunner of the helicopter);
- in 1910, Aurel Vlaicu realizes the first flight using an aircraft of Romanian construction;
- in 1910, Henri Coandă realizes the first flight using an airplane with no propeller;
- in 1920, George de Bothezat realizes a helicopter and brings many contributions to the flight stability [41].

Fig. 1.2. *Traian Vuia and Clement Ader near the Vuia I airplane* (1906)

Unfortunately, at that time, the Romanian authorities showed a *"totally lack of interest"* in the aerospace area [03], except Aurel Vlaicu. All others worldwide recognized pioneers put in practice their accomplishments in countries where they succeed to find financial support. Even though the Romanian inventors started developing safer and better airplanes in the same period with the others countries from Europe, they didn't resist in time, as producers who succeed to develop a durable business in Romania. If the names as Wright, Potez, Breguet, Curtiss, Fokker, Heinkel, Dornier or Junkers have lasted till our days, Romania doesn't know such a success. The aviation is a very difficult business sector, especially in countries where the economy isn't so developed. In these countries there were small business or state owned companies which, during the time, there were privatized or nationalized, respectively, meaning that aviation manufacturing business involve big risks. Some countries created national industries only because the acquisition of foreign aircrafts was far more than their financial possibilities. The Romanian aerospace industry makes no exception, in the present time, the big producers limiting their activities or reorienting them on the aerospace structures development and manufacturing.

1.3.2 Priorities of the worldwide aviation

In more than a hundred years since the controlled flight, the aviation reached the level maturity, the actual trends being the improvement of the so far technical achievements. For instance, related to the structure, aerodynamics or the engines, after 1980, there is no significant news. It was calculated that, because of the big passengers' number / goods and its constant increase, such savings of $5 \div 10$ % represent important amounts for the airlines.

As of priorities, the followings can be identified:
- the reduction of the fuel consumption using the improved engines and lightweight structures;
- the weight saving of structures by the transition to the light materials and structural optimization;
- the reduction of the maintenance costs by introduction of the real time structural health monitoring;
- the increase of the comfort level by the ergonomics and computerization;
- more environmental friendly aircrafts, with noise reduction and less pollutant manufacturing processes;
- the reduction of the production and development costs.

The governments agreed with the following of these directions by the allocation of funds or incentives both for the companies and the industrial clusters. Although, the implementation of new flight configurations or alternative propulsion techniques is less likely in the near future, different programs as Clean Sky, financed by the Horizon 2020 program, encourage also new aircraft configurations, stimulating the innovation and the development of alternate solutions to the current ones.

1.4. THE MAIN COMPONENTS OF AN AIRCRAFT

1.4.1 General

Generally, an aircraft consists of structure, engines, installations and on board systems. The domain of the structure engineer is the aircraft structure and the integration of the main components like engines, landing gear, hydraulics, electrical systems and avionics. The main components of an aircraft are shown in Figure 1.3.

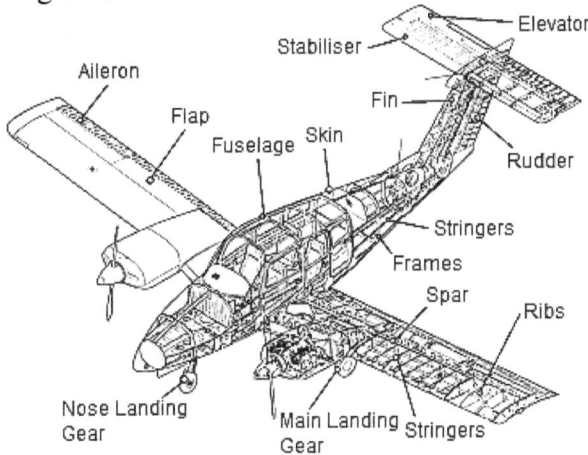

Fig. 1.3. *The main components elements of an aircraft* [55]

The state of the art solution for an aircraft consist in one stiffened wing including the fuel tank, a fuselage for the payload, and an empennage for stabilizing and controlling the flight altitude in the symmetry and horizontal planes. From the structural point of view, this configuration isn't the optimal because the lift is produced in a different area than in the one where the payload is found [49]. The optimum solution is the flying wing, but researches made in the 80's indicate an unpractical concept under a payload of 500 t, which exceeds by a lot even the biggest aircraft that exist today [49].

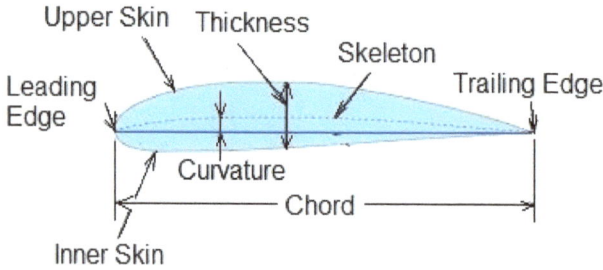
Fig. 1.4. *The components of an airfoil*

The main stiffened element is the wing, a very important component for the aircraft performances. The main elements of the airfoil (wing section) are illustrated in Figure 1.4. By the structural point of view, the wing is a beam embedded in the fuselage, presenting one or more spars. Due to the fact that the lift is the result of pressure differences from the lower and the upper skins, the structural connection between the wing skin and spars is realized by a system of ribs. The ribs keep the airfoil shape under aerodynamic loads; between ribs the role is taken by the longitudinal stringers. The empennage and command surfaces (ailerons, flaps, slats, elevator and rudder) have a similar structure.

The fuselage has the same structure as the wing, the ribs being replaced by the frames. Both the wing ribs and fuselage frames may be of light construction (preserve the swept shape and withstand only to aerodynamic loads) or heavy loaded (contain the mounting points for engines, wing, landing gear, etc.).

1.4.2 The classification according the failure impact

The basic requirement of an aircraft structure is to withstand to the load cases requested by the certification regulations, for the lightest possible structure. The operating safety means the correct finding of the maximum loads during operational flight (limit forces) and the design of a structure that will withstand to higher efforts than the ones during flight (ultimate forces). The ratio between the ultimate forces and the limit forces represents the reserve factor. Because a higher reserve factor leads to a robust structure (a heavy structure), a ranking of the structure elements was elaborated. This ranking has been made related to the impact of that element over the airworthiness of the aircraft (the capacity of fully controlled flight and landing in defined operating conditions). The elements of the airplane structure are classified as follows:

- vital elements (the primary structure) – their failure leads to the loss of aircraft structural integrity and a catastrophic failure (for instance: wing box, fuselage, stabilizer);
- structural significant elements (secondary structure) – their failure jeopardize the normal continuation of the flight, thus the emergency landing is a must (for example: the control surfaces);
- elements of tertiary structure – their failure doesn't impede the flight continuation, but their fix after landing is requested (for instance: fairings, engine cowlings, landing gear doors).

Using different reserve factors, a weight optimized structure is obtained, the robust ones being only the vital elements, the others having a slender construction.

1.4.3 Materials

From its beginnings, the aviation used the wood and fabric as primary materials, the metallic elements being used only for the engine, tensile wires and brackets. Later, steel, Aluminum alloys and stainless steel were used. By increasing the speed, the titanium started to be used; after 1960, the composite materials were gradually used, in different combinations and manufacturing techniques.

At the beginning of the aviation, the wood was the first material used for the airframe components [30]. At that time, the wood was the only material to build surfaces light enough to fly, and also to withstand the flight maneuvers loads [30]. Moreover, the excellent manufacturing know-how, repairing techniques and the reduced costs are added [30].

According to [23], the biplane structures were used many years after the aviation beginning and there were associated with lightweight materials as bamboo and spruce (Fig. 1.5).

Reference [23] consider that *"although, technically the wood was considered as outdated and the metal as innovative, the aviation was the area where these materials were used together in a complex and interesting way"*. Even though it is considered that the metallic construction leaded to the major changes in design and the transition to the aerodynamics shapes and monocoque structure that appeared only after those new ideas were tested on aircraft with wood structure [Jakab, 1999]. As a proof, in [Hoff, 1946] it was considered that the transition from the wood to metal was done without taking into account the technological and mechanical particularities of the last one (only by the material replace).

Although, the technical advantages of the metal were not very well known, since 1925 – 1930 there were communities of producers who were against the use of wood in aircraft construction, considering the metallic aircraft as *"superior and advanced"* [30].

Aluminum started to become available at reasonable prices only after a decade [30] (before 1900, its price was comparable with the silver price) [78].

Fig.1.5 *Wing from spruce and bamboo* [F01]

In the Table 1.1, there are illustrated the values for ultimate stress, density and the structural efficiency for a set of materials used in the aerospace industry.

Wood, duralumin and/or composite materials are used for spars, stringers, skins and for other primary structure's elements.

The duralumin is the material which was widespread used, with the biggest share relative to other materials, being used for a long period of time. In [47], it is stated that it is the material with the most important role in the aircraft production.

The textile skins (made by fabric impregnated with cellulose acetate and painted [64]) were used for the stressed surfaces and the fuselage covering (except the engine area). The fabric was used as being the lightest, but the main disadvantages were related to the reduced durability and the very limited possibilities of fireproofing [71]. Steel was used for the joining components (fittings), welded frames, landing gear, wearing parts or bushings (together with brass). The stainless steel is used for corrosion exposed parts, in the wearing areas, and leading edges.

Table 1.1 [43], [01]

Material	State	Ultimate stress (F_{TU}) [daN/mm^2]	Density (ρ) [kg/m^3]	Structural efficiency (F_{TU}/ρ)
Spruce wood		100	450	0.22
Birch plywood	Bakelite glue	75	800	0.09
Duralumin	2024 T4	476	2710	0.18
	7075 T6	546/2710	2710	0.20
Titanium	6A1-4V annealed	938	4507	0.21
Steel	4038	1260	7800	0.16
Magnesium	AZ31B-H24	280	1740	0.16
Glass fiber	with epoxy resin	560	1761	0.32
Carbon fiber	with epoxy resin	1190	1517	0.78

Titanium alloys are used for the firewalls, engine fittings, assembly elements and other thermal stressed parts. Magnesium has a limited use due to the risk of the self-ignition. It is used only in alloys, for the gears housings.

Fig. 1.6 *Main Gear Housing made by cast magnesium alloy* [F02]

From the example of the wood to metal transition, it may be stated that the transition to composite was firstly done only by the replacement of material. A second stage is the one of the adaptation of the components design to the possibilities introduced by the new composites manufacturing technologies. On this direction, there were already made important steps, as:

- raw materials profiles made from composites (similar to the laminates or metal extrusions) (Fig. 1.7, a);
- parts manufactured by thermoforming (Fig. 1.7, b);
- fuselage composite panels replacing the fabric with prepreg composite straps (applied by a robotic arm - Fig. 1.8, a);
- monolithic fuselage barrels to exclude the longitudinal joints of the fuselage panels (i. e. Boeing B787) (Fig. 1.8, b);
- new assembly methods (integrated or stitched parts).

a)   b)

Fig. 1.7 *Carbon fiber parts*
a) *Profiles [F03] b) Thermoformed part* [F04]

If the transition from wood to metal lasted 25 years, the transition from metal to composite started in 1970 and it's not over yet. This is due to the fact that in a single composite part, a big number of combinations of materials, resins, number and orientations of plies are possible. Thus, understanding the properties and the behavior of this polytrophic material needs a long time. By comparing it to the wood – metal transition, the transition to composite presents some different particularities:

- the metal was introduced from the beginning in the loading areas, while the composite was introduced many years only in nonstructural areas (fairings, command surfaces, etc.); as an exception, were the helicopter blades, pioneering the composites in the primary structure;
- wood was definitely removed from the airplane structure; in the most advanced composite structures, the metal represents at least 30 %;
- excepting the gliders, after the transition to metal, the wood airplanes were not built anymore. The metal remains the main material in the development of most aircrafts. Its behavior at impact couldn't be replaced yet.

As similarities between wood and the composite materials, the followings may be noted:

- no one is homogeneous and isotropic;
- no one presents the plastic range before total failure;
- the plywood is a composite, too;
- both present low bearing stress;
- the manufacturing doesn't require high qualified personnel.

Although the material has changed, the most widespread structural layout, consisting in stressed skin, frames and stringers was kept, with the technological updates for the parts and assemblies manufacturing, specific for each material (Fig. 1.9).

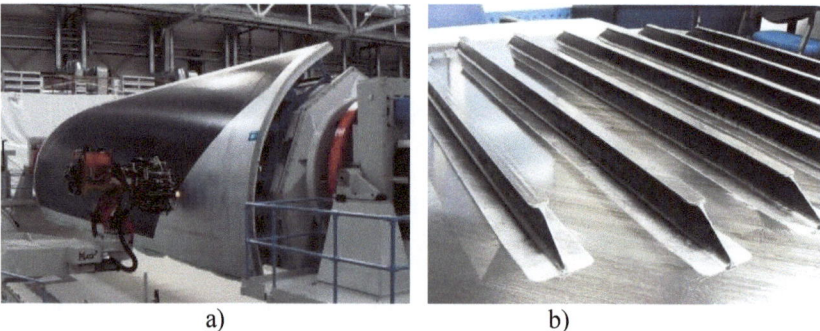

a)   b)

Fig. 1.8 a) *Fuselage panel built by overlapping bands* [F05]; b) *Panel with integrated stiffeners* [F06]

Table 1.2 Assessment of structural materials [01], [35], [75], [64]

Material	Advantages	Disadvantages
Wood	Small price Easy and cheap manufacturing Easy repairing Good structural efficiency	Variable properties Anisotropy Poor crushing strength Water absorption Dimensional instability Brittle material Delamination (plywood) Exposed at decay and fungus Dependence by adhesive (bonding layer fracture)
Linen fabric or mercerized cotton	Small price Very low density	Wear under the sun and water action Flammable Not withstand to compression loads Low penetration strength Needs retreatment at every 3 − 5 months
Duralumin	Good machinability Wide range of alloys	Easy corrosion Poor behavior at the fire
Steel	Small price Good mechanical properties Good contact strength Weldable	High density Easy corrosion Low structural efficiency
Titanium	Good behavior at high temperatures Don't need surface protection	Poor machinability Expensive
Composite materials	Good structural efficiency No corrosion Can be modeled in complex shapes	Delamination risk Hygroscopic Difficult to repair Difficult at NDT inspection Expensive and pollutant technology

It may be concluded that each material was the most appropriate for its period:

- wood, for the low speed and lightweight, well fitted with the beginning of the aviation;
- metal, for the increased speed, payload and durability; without this material it would have been impossible to conquer the outer space;
- composites, for the overall improvement of the structural parameters, a better distribution and orientation of the material according to loads, easier manufacturing of complex surfaces, and increase of the life time.

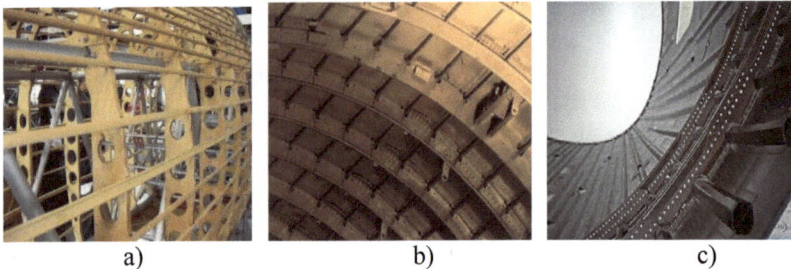

Fig. 1.9 *Structures with frames and stringers, made from:*
a) *wood*; b) *duralumin (Airbus A380)*; c) *carbon fiber (Airbus A350 XWB)*

Most likely, if the composites manufacturing technology would have been developed 50 years before, the producers would not have been able to exploit the opportunities offered by these materials.

1.5. THE LIGHTWEIGHT CONCEPT

The lightweight of the flying machine represented from the beginnings of aviation a critical requirement. If the aircraft payload is about 40% from the maximum take-off weight, results that an increase of 20 % of the own weight of the flying machine leads to a 30 % decrease of the payload. Maintaining the payload can be done only by the reduction of the quantity of fuel, leading to range or engine power reduction. Unfortunately, the aviation history has many cancelled programs, many prototypes being too heavy and not compliant with the technical specifications, often leading to the orders being cancelled and the whole development program ceased.

Designers who proposed extreme lightweight structures faced problems related to the corrosion or fatigue behavior. It follows that it's very difficult to realize the compromise between the lightest possible structure, strong enough to withstand the loads specified by regulations. This requires gathering of a certain level of knowledge based on few aircraft generations, combined with the feedback from operational field of every fleet, that means at least 50 years of experience. For this reason, the main commercial and military aircrafts operating after the WWII were developed by such producers.

1.5.1 General

According to reference [59], "*a considerable effect over the weight reduction can be obtained only by a careful study and by summing of all the small reserves of weight*", and "*it's hard to find an airplane with no possibilities to save weight, fact explained by its natural weight reserves.*"

In [27] it stated that "*the lightweight of the assembly can be obtained only if each component reached a minimum; the optimum of the individual parts contributes to the optimum of the entire assembly*". Therefore, Hertel recommends searching for weight reduction resources even in the smallest parts.

In [12], it is mentioned that "*starting from 1970, from one year to another, the aircrafts are with 2% more efficient in fuel consumption; this fact is due to better engines (2/3) and to the lighter structures (1/3).*"

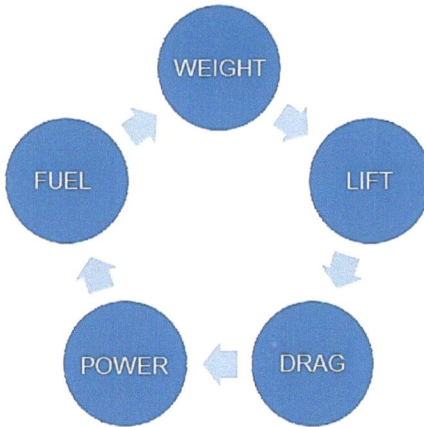

Fig. 1.10 *The structural weight reduction cycle* [17]

Many authors insisted on the cycle of the structural weight reduction (Fig. 1.10). Because a lighter structure needs a lighter wing and a smaller engine (with lower fuel consumption), a reduction of the structure weight have important implications over the maximum take-off weight. The structure weight of an airplane is between 24 – 35 % of the maximum take-off weight [61].

According to [27], is stated that a variation of the structural weight with 10 % can lead to a variation of the maximum take-off weight with at least 40 %. In literature it is also presented the financial implications of the structural weight reduction. These, computed on the operational period of a fleet, lead to savings of millions of Euros on every percent of reduced weight.

On the other hand, in [27] one affirms that the design which has as single objective to withstand the operational loads is not enough for a safe solution. For instance, a poor quality of the aerodynamic surfaces (wavings deeper than 0.25 mm/m) will lead to the decrease of the aerodynamic qualities, involving more power from the engine, therefore a weight increase [27]. Also, the weight reduction can lead to a too elastic structure, with poor behavior under the aero elastic phenomena (flutter and buffeting) [27]. Therefore, the lightweight design can be reformulated as an assembly of requirements: lightweight/surface quality/stiffness.

1.5.2 Traditional approach

In the traditional approach, from the early stages of the project, it is insisted on the maximum accuracy estimation of the aircraft weight [43], [59], [61], [42], [15]; exceeding the target weight involves the payload or range reduction, fact that can affect the entire project or can lead to the producer bankruptcy.

The lightweight structure is controlled by politics or design methodologies very strictly monitored. After the prototype construction, a pre-serial aircraft is usually subjected to one or two design iterations of weight saving.

The methods of aircraft weight reduction consist in:
- the high accuracy calculus of the operational loads;
- the correction of the analytical calculated loads with those obtained after the flight tests;
- the correction of the fatigue loads after the data acquisition from the entire fleet behavior after 50 – 80 % time from the operational period;
- the experimental determination, of the capable force for joints and fasteners;
- the experimental determination of the geometrical parameters of the riveting (i.e. the pitch and the edge distance);
- the redundant control of the structural weight (by weight engineers; one method is to compare the developing aircraft with other aircrafts from the same category);
- the development of materials with minimum density;
- the use of the dedicated materials according to the predominant loading on that area (tensile, buckling, bending, corrosion, fatigue, impact, vibrations, thermal stress);
- the alternate design (generation of different design concepts, and the lightest one will be selected for detail design);
- the hybrid design (the simultaneous use of different technologies – riveting/ welding/ bonding/ composite for an efficient mix of their advantages);

- research and innovation, etc.

A weight reduction of 5 – 8 % for each design iteration is considered feasible [15]. During an aircraft development program, the airframe may obtain a weight saving of 15 – 20 %. Such a program may be ordered by an operator who owns a large fleet also within an upgrading program which targets other aspects (i.e. new engines, avionics and weapons).

1.5.3 Design principles

The design teams define for each new aircraft in development a design handbook, where is a chapter dedicated to the lightweight principles. Among these, it identifies [15]:

- keeping of the secondary structure at a minimum;
- reducing the number of joints;
- reducing the number of fasteners;
- keeping only joints subjected to shear loads, and avoid bending moments;
- the minimum edge distance (from fasteners to parts margins)
- the minimum wall thickness for all parts;
- the minimum bending radius for the sheet metal parts;
- keeping the parts volume to the minimum using milled pockets, radii and chamfers;
- the removal of the excess of shims, fillers, sealants and spacers;
- keeping at minimum the number of parts;
- replacing of the adjacent parts with a single one;
- manufacturing of the webs and skins of variable wall thicknesses.

The highest weight reduction is done by the use of the torsion box which behaves better under combined loads. This solution is found in all the stressed structure components.

Fig. 1.11 *Lightweight details: honeycomb structure, sheet metal webs with holes (Airbus A350), parts with milled pockets (Airbus A400M)*

The torsion box is suitable for closed volumes, in practical design, one try to maintain a minimum ratio between the openings areas for (windows/ doors/ hatches) and the rest of the skin.

Another important aspect of lightweight is the detail design. The state of the art solutions are the sandwich structures and structures with lightening holes (Fig. 1.11). Applied especially for large parts, this involves the use of some standardized methods which are technologically feasible and have a good behavior under vibrations and fatigue loads. Besides the holes or milled pockets for weight reduction, there are methodologies to define the joining of structural elements (frames, stringers, ribs, skin) or the structural subassemblies (stiffened panels, doors or control surfaces).

1.5.4 Trends in lightweight design

The lightweight design still represents one of the main research priorities for the current aircrafts. From aerodynamics and structural arrangement point of view, convenient solutions, together with the researches for better engines were reached, but the lightweight design represents one of the directions where a revolution is not expected, but only improvements and characteristics adjustments. For the modern aircrafts, the following trends are identified [16]:

- the use of the sandwich panels with foam instead of honeycomb (especially at the fairings);
- machined frames instead of riveted frames (reduction of the number of joints, the parts and fasteners);
- replacing riveting by welding (for the fuselage panels stringers, in the areas which are not subjected to fatigue loads and corrosion);
- the widespread replacing of the steel fasteners with the titanium fasteners (even if titanium is three times more expensive than steel);
- replacing of the steel parts or of high loaded aluminum parts with titanium parts;
- building of composite laminated parts for longerons and profiles (Fig. 1.12, a);
- using composite thermoformed parts (similar process with hot forming of the deep drawn parts) for the structural parts or mounts (brackets, corner angles, intercostal parts, closing parts) (Fig. 1.12, b);
- the transition of the primary structure to composites for the large aircrafts (Boeing B787, Airbus A350).

Within a study targeting a number of 65 passengers' aircrafts, it was defined a structural weight index, as follows [15]:

$$i = \frac{M \times 10^6}{N \times R \times S^2}$$

where:

M – maximum take-off weight (*kg*);
N – number of passengers;
R – range (*km*);
S – cruise speed (*km/h*).

The results are presented in Figure 1.13, which doesn't include the first passenger airplanes - Vickers Vimy (1917) and Farman F60 (1919) - because of the very high values of the structural weight index (44 and 59, respectively). The most significant progress is observed between 1920 and 1940, due to the tripling of the range and the doubling of the cruise speed. After 1960, a stabilization of the performances and a lower progress rate is noticed.

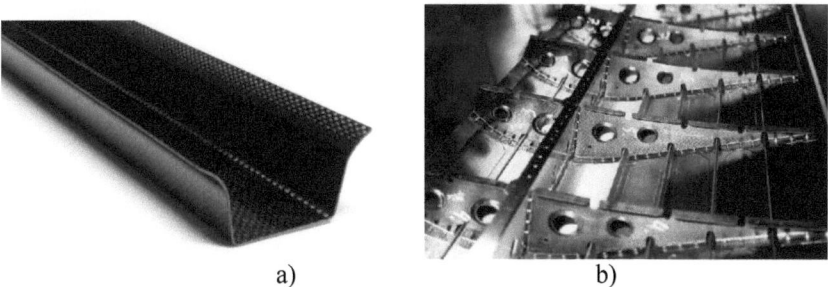

a)   b)

Fig. 1.12 *Thermoformed parts:* a) *Longeron;* b) *Ribs (Dornier Do 328) [F07]*

From Figure 1.13, some conclusions may be drawn:
- in time, the structural weight is use more and more effective;
- the large aircrafts are the most effective;
- on the top are situated the Airbus A350, Boeing B747-8I, A380 and B777;
- the general trend of the last 50 years shows the possibility of improvements with a slower rate by comparing with the years 1960 – 1980;
- a bottleneck of improvements is expected around 2030 – 2040;
- the trend of the best aircrafts of the last 50 years indicates improvements in effectiveness of about 15% for 2030, relative to 2010.

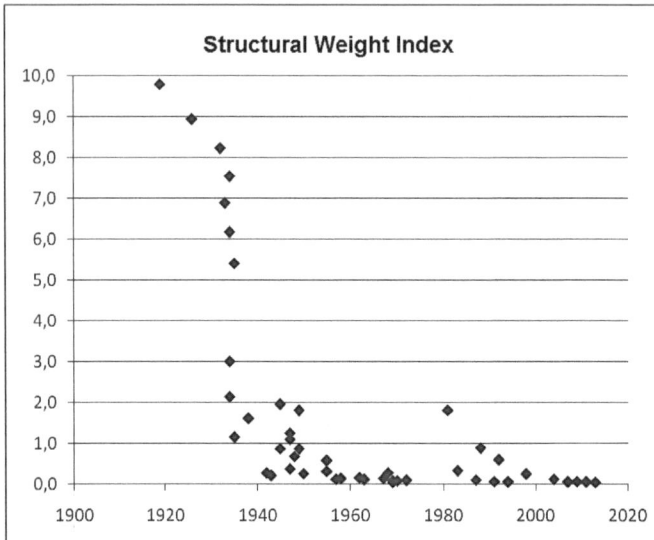

Fig. 1.13 *The evolution of the structural weight index for the transport aircraft*

It cannot be ignored the trend to eliminate the pilots, where this is possible (low payload aircrafts) or necessary (especially the military and long mission aircrafts) by using the drones. This trend became very interesting, from the defense departments to amateurs, new models existing in a very big number and diversity.

Besides the aerospace structure producers, efforts of weight reduction at the equipment producers can be observed, especially for the ones which have big percent from the total weight of the aircraft. Thus, some progresses are noticed in areas like engines, landing gear, seats, fitting, hydraulics and wirings.

It has to be mention that, due to the extremely use of the lightweight design politics, in some design teams, the old fashioned stress engineers, consider that the aircraft aren't equally safe as the ones from 80's; this issue doesn't have to worry anyone as long as the certification process proves by analytic, experimental and flight test means that the aircraft is fully compliant with the airworthiness requirements.

1.5.5 The role of structural optimization

A project engineer designs is often, based on its own experience, the project being the result of certain criteria: technological simplicity, minimum manufacturing costs and good corrosion and fatigue operational behavior. Regarding the parts shaping, these are generated relative to the prevailing loads: hollow cylinder for torsion and buckling, *I* profile for bending and torsion box for the combined loads. For large structures, the latticed beam is considered to be the most economical. Its proportions and the triangulated cells represent well-known patterns.

In aviation, the most used structure is the monocoque, employing frames (parallel with the cross section) and stringers (parallel with the generatrix of the aerodynamic surface). The structural arrangement, the pitch of frames and stringers, the wall thicknesses are often chosen by experience, the calculus being done only for checking purposes.

The load cases are in a big number, their loads being a combination of forces and moments; the difference of pressure is added for the aircrafts flying over 5000 m altitude. The designer has to create a structure which must have a good behavior for all loads cases that makes almost impossible to obtain an optimized structure in the traditional design approach.

In the last years, different techniques of structural optimization started to be used, allowing the design of parts that withstand at the same loads, for a lighter structure.

The commercial software solutions for structural optimization offer on a simplified model of the parts, the flow of stresses, qualitative indicating to the designer where stiffeners are needed (formed or added – from the topography optimization) and/ or wall thickness has to be increased (for the topology optimization). In further optimization iterations, values recommended for the wall thicknesses may be obtained, a process that is very useful for dimensioning.

The final dimensions may be determined in a small number of iterations (2 - 3) of design/check by finite elements analysis, by the final adjustment of the wall thicknesses or radii values, where the stresses exceed the allowable limit load. In Figures 1.14 and 1.15 are illustrated the main stages for a project targeting the structural optimization of a fitting (to attach a pilot seat to the cockpit floor).

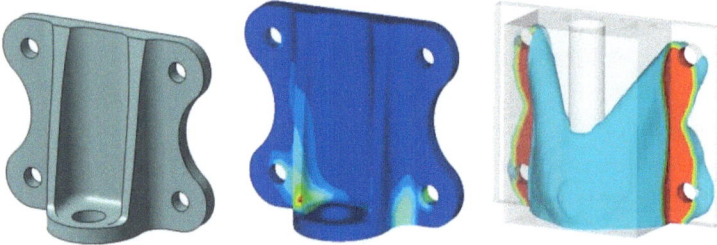

Fig. 1.14 *The initial geometry, the stress gradient and the proposed solution of topological optimization for a fitting*

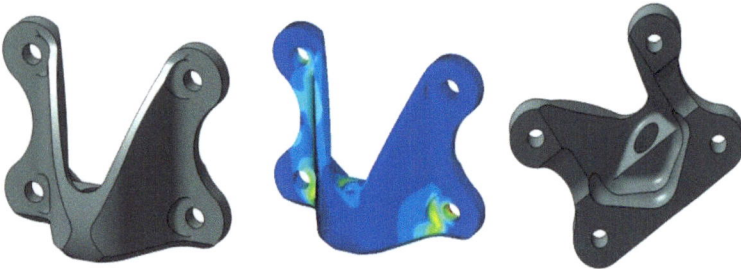

Fig. 1.15 *The geometry and stress gradient of the re-designed fitting*

It can be observed that the re-designed part presents a more widespread Von Mises stress gradient. The maximum of the Von Mises stress is only 50% comparative to the initial design, and the displacements are three times lower, for a weight reduction of 13 %.

The conclusions of this project were:
- the optimized design is superior to the initial one by strength, stiffness and weight;
- the re-design of the part after the topological optimization results is a difficult and time-consuming activity;
- the final shape of the part is complex, requiring a five axis milling machine;

- the topological optimization and the final design need top class software solutions, high qualified personnel and big resources of time, which involve high expenses.

Bombardier Canada reported 10 % weight savings for a wing rib [07] after an assessment between the classical design and the one that uses structural optimization solutions (Fig. 1.16), and according to [10], it was reported a reduction of 19 % for the arm of the passengers door of Fairchild Dornier 728.

In 2006, Airbus reported that using the topology optimization weight savings between 20% for A350 and 40% for A380 were obtained [57].

Moreover, the topology optimization came up with very interesting conclusions against the superiority of the monocoque structure relative to the latticed beam. Thus, a monocoque hybrid structure, with a lower number of frames, equipped with internal bracing and shear bearing diaphragm walls, presented lower stresses for a reduced structural weight [57].

If, since the ancient times, the architect was the one who shaped and dimensioned the building, and was responsible for its strength, in time, it appeared the need of design and calculation engineers. The trends from topological optimization area require a close collaboration between the designer and the calculation engineer. If nowadays the calculation engineer checks or optimizes the parts created by the designer, in the future, the designer will have to generate parts which already respect the stress flow in order to reduce the number of iterations and time required to reach the optimum design.

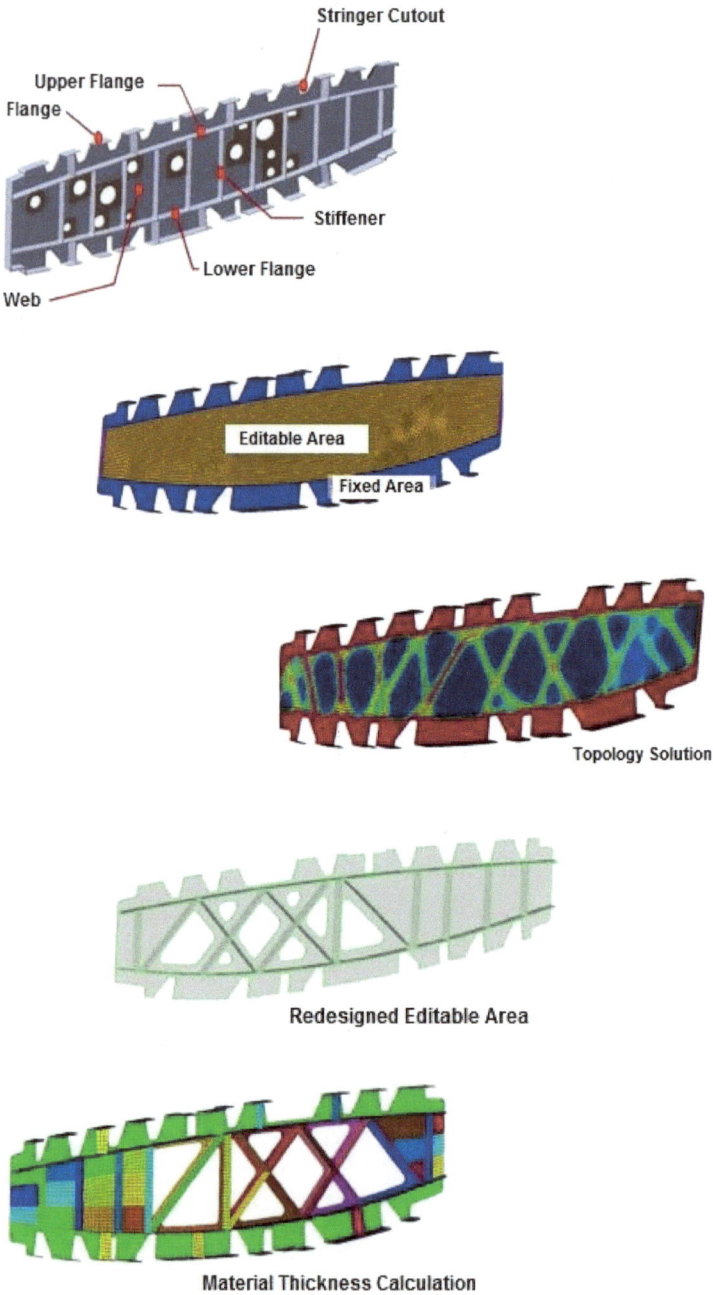

Fig. 1.16 *The stages of the topology optimization for a wing rib* [07]

The needs of the present time indicate the return of the need for both design and calculations to be made by the same person. This may be possible by a customized engineering and appropriated software training. This trend was understood by the CAD (Computer Aided Design) software producers, by adding finite element calculation modules to the basic CAD software; due to results under expectations, this strategy was cancelled, CAD and CAE (Computer Aided Engineering) software being further developed as separate solutions. It can be concluded that the topology optimization is about to bring a revolution in the development of structures. In the following years, there are to be expected:

- more efficient and also affordable software solutions for structural optimization;
- a new approach in engineering universities of mechanical design, based on shapes resulted from the structural optimization – organic/ hollow shapes, taking into account the stress flow;
- the development / upgrading of a new technology as Additive Layer Manufacturing or 3D Printing, to allow the manufacturing of the complex shaped parts obtained from the structural optimization;
- the development of appropriate quality control methods for the parts with complex shapes.

If in practical design, the linkage between the project and the manufacturer is made by the manufacturing drawing, the complex shaped parts will become impossible to be dimensioned; therefore this role is expected to be taken by a new generation of virtual 3D model of part. This process is already applied, but only on a small scale because its limitations, the most challenging being the massive orientation to low cost products.

1.6. THE REQUIREMENTS OF THE AIRPLANE STRUCTURE

Among the aircraft requirements, the structural integrity has a special place. The most important characteristic of the aircraft is the airworthiness [12], defined by Cambridge Aerospace Dictionary as: *"the ability of operating during flight, in all the possible conditions and in all the estimated circumstances for which the aircraft was designated"*.

The structural integrity is a direct subsequent of the airworthiness requirement. The objective of the aerospace engineer is to produce the lightest structure which can provide the airworthiness [12]. The aircraft fly at higher altitudes and speeds by comparison with other vehicles, for this reason, in most cases, a structural failure will lead to a catastrophic event. The big variation of temperature in every flight (up to 120 Celsius degrees) and the long operational life (in some cases, over 40 years) are beginning to give an overview of the operating conditions. J. D. Anderson stated: *"although the aerodynamics is good, the propulsion is strong and the flight dynamics is spectacular, if the aircraft lose its integrity, then everything is worthless"* [02].

Fig. 1.17 *Drop test of a fuselage barrel* [F08]

Although, there are known aircrafts with modest aerodynamic performances, under-powered or too heavy, no one can fly without a structure 100% safe. It may be stated that the performance of a structure doesn't consist in the level of safety (which is an on/off issue), but in a structure as light as possible, withstanding all the certification regulations loads.

The loads of the aircraft structure are flight and landing loads. A special request is for the behavior at impact with the landing strap, the basic condition being to absorb the shock in order to maintain the structure integrity for the passengers' protection. In Figure 1.17 is illustrated the result of a drop test on a fuselage barrel where the structure behavior after an impact with a speed of 9 *m/s* was investigated.

Fig. 1.18 *Fuselage section folding, keeping the inner volume* [F09]

Fig. 1.19 *The burned fuselage due to the ground impact fire* [F10]

In Figure 1.18, there is shown the fuselage structure folded only in the area located under the passengers' floor (cargo area), the passengers' compartment keeping its own volume. It may be observed the similarity of the structure and seats behavior in the test and real cases.

Other implication of the impact with the ground is the large fuel amount with high octane value (approximately 200 t) which is spreading, the risk of ignition and a big fire being imminent. This, together with a poor behavior of duralumin in the fire conditions, leads to the structure failure in just a few minutes. Fig. 1.19 presents the accident of the same aircraft, a (Boeing B777, Asiana Airlines Company). The aircraft met the conditions for both scenarios: impact with ground, keeping the volume of the passengers' compartment and a structure failure due to the fire. It may be concluded that the second scenario has more critical consequences for the passengers.

Moreover, the structural design does not have the only purpose of aircraft integrity. This has a special role in the flight dynamics [12]. A too elastic aircraft can lead to the occurrence of the aero elastic divergence phenomena like flutter (oscillations along the wing chord) or buffeting (the wing oscillations in vertical plan), leading to catastrophic effects for the structural integrity.

Another aspect is related to the fact that the aerodynamic forces have very low values (below 1 daN/cm^2), for this reason the wing structure cannot be a monoblock type, the solution being the thin-walled structures [49]. The optimum structure is reached when thinner and thinner components are used toward the areas withstanding the aerodynamic loads. Thus, a compromise is obtained, the aerodynamic areas being thin walled, while the heavy loaded areas having robust elements, between them, a transition structure being needed.

1.6.1 Loads on the aircraft structure

The loads an aircraft has to withstand through its operational life are specified by the civil regulations (FAR, JAR, BCAR, EASA) or military rules (MIL-A-887). According to references [44], [37], these loads are as follows:
- ground loads (taxiing, roll, landing);
- flight maneuvering loads (the upper limits are identical with the human body accelerations limits [49]);
- gust loads;

- engine loads;
- mission related loads (for military and general aviation aircrafts);
- loads induced by the payload (freighter aircraft or external attached loads);
- pressurization loads (only for the aircraft flying over 5000 m height).
  The inner/ outer pressure difference may be up to 5.6 t/m^2 for the pressurized fuselage [75].

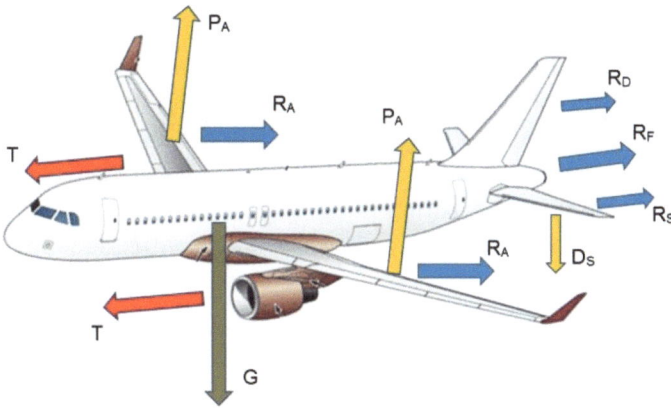

Fig. 1.20 *The forces balance for the horizontal flight* [F10]

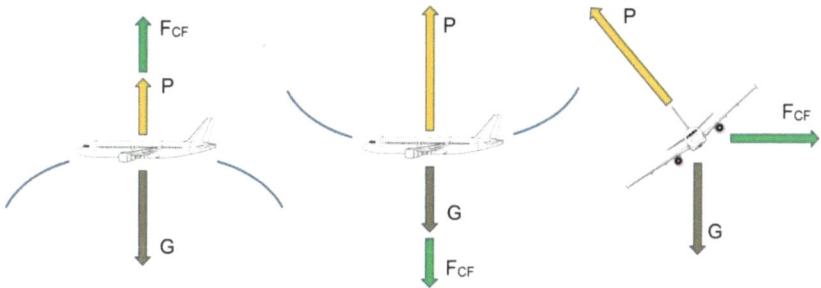

Fig. 1.21 *The forces balance in vertical and horizontal manoeuvers*

In figure 1.20 there is presented the force balance of the horizontal flight of an aircraft with classic configuration - wing with stabilizer (T – the engines thrust, P – the lift, D – negative lift, R – the aerodynamic drag). The forces balance is realized so that, while approaching the stall speed (the speed when lift become lower than weight), the aircraft will pitch down in order to increase the speed and recovery the lift. It is observed that the stabilizer has negative lift, only in the canard configuration having positive lift (with nose stabilizer).

In Figure 1.21, there are shown the forces for the maneuvers in vertical plan (negative and positive resource), respectively, at the steering in horizontal plan. It can be observed that, because of the centrifugal force, the structure loads increase, for the case of military or acrobatics aircrafts reaching to the values up to 9 *g*.

For calculations purposes, for each major component, the loads are isolated (the rigid body diagram method). For instance, the wing and the empennages are considered being embedded on the fuselage. In its turn, the fuselage is considered embedded in the wing and empennage fittings for flight maneuvers. For ground stress calculations, fuselage is considered embedded on landing gear fittings.

Forced from load cases are converted in static loads (tensile, compression, bending and torsion), dynamic, stability loss (buckling) and fatigue.

The flight safety requirements of the aircraft structure are presented according to [51]:

- the stresses values in the components must be lower that the ultimate stresses;
- the elastic deformations due to the accelerations in flight maneuvers and the kinetic heating must not exceed the specified values;
- cracks and fractures in the parts due the fatigue loading, must not occur;
- the aircraft must be rigid enough not to modify to its shape; otherwise, the aerodynamic characteristics will be changed and the aero elastic phenomena may occur;
- the joints between the major components may present small local deformations, without affecting the aircraft strength and stiffness.

The maximum forces estimated to occur during the operational life are named limit loads. These are only peak loads, the probability of occurring during the life time of the aircraft being very small (one time maximum). For the limit loads, the aircraft must present deformations which cannot disturb the accurate functioning of all systems on board; these deformations have to remain in the elastic range [44].

The limit loads multiplied by the safety factor (at least 1.5) become the ultimate load; the aircraft is calculated to the ultimate load [06, 44]. The ultimate loads must be supported by the structure for at least three second with no failure (remnant displacement are allowed).

1.6.2 Other certification requirements

The certification requirements of the structure are not limited to the mechanical behavior under loads. To these, other requirements are available, as follows:

- requests related to the aerodynamic surfaces quality (barrels and panels flushness / gaps, the surface waving, roughness, continuity of paint);
- corrosion strength;
- durability;
- fire proof of all combinations of materials (including coatings, paining, sealing, shims);
- airtightness (for pressurized structures);
- buoyancy (for the emergency landing);
- the escaping/ evacuation from the damaged airframe;
- inspectable;
- maintainability.

The certification process of an aircraft is long and expensive. The definition of the certification requirements is done in the early stages of aircraft definition – from the elaboration of the preliminary technical specification. These are made by the producer and the governmental aeronautic authority. The role of a certification process is to demonstrate the fully compliance of the prototype to the requirements of the certification regulations. After obtaining of the certification for the prototype, every series aircraft has to prove the conformity with the manufacturing drawings and technical specification by the quality inspection and flight tests.

Also, each modification or repair must be documented, and according to the intervention level the project needs, an update from the stress calculations and the risk analysis in order to demonstrate that the aircraft airworthiness isn't affected. Due to the small manufacturing series, the situation when there are no two identical aircrafts will appear which requires a very strict configuration management.

CHAPTER 2

The evolution of the aerostructures

The flight was possible only when a minimum number of conditions were accomplished: an adequate aerodynamic configuration, a source of sufficient power, a stiff and lightweight structure, and appropriate controls to provide the flying machine's stability and maneuvering during flight. The flight safety was an unknown and for this reason, many aviation pioneers lost their lives without being aware of the dangers to which they are exposing. Although the basic principles of flight mechanics were known from the beginning of the 19th century, the human has succeeded to fly only a century later, the last tests before this event being failed due to some fragile structures.

The present chapter is referred to the primary aircraft structure (structure whose failure leads to the loss of the aircraft integrity), the control surfaces, the installation of engines and the landing gear arrangement. In parallel, there are analyzed the evolution of the structural concepts, materials used and manufacturing technologies. The chapter doesn't deal with the installation of the onboard equipment, the military installations, avionics, etc.

2.1. THE EARLY PERIOD

In 1810, Sir George Cayley made the first sketch of a glider consisting of fuselage, wing and empennage (Fig. 2.1 a, b). Also Cayley introduced the wing curvature, the wing dihedral angle, the aircraft balance and the equilibrium of forces in the horizontal flight [09]. He built a glider in 1853, and he realized the first flight with human onboard [71].

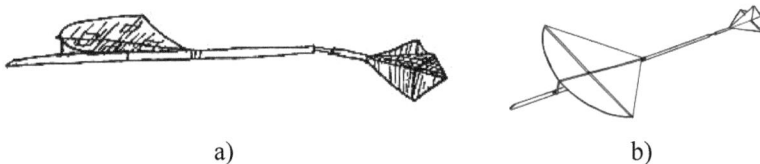

a)   b)

Fig. 2.1 a) *Sketch of a glider realized by George Cayley in 1804* [41];
b) *Axonometric of the Cayley's glider* [F11]

In 1842, Henson realized a project of a glider that had the configuration of an aircraft, where it may be recognized the wing spar, the ribs and also tensile wires system aimed to provide the wing stiffness (Fig. 2.2). According to some authors, the Henson's glider is the first modern approach of the aircraft structure.

In 1894, Frederick W. Lanchester built and tested Aerodon (length of 1.8 *m*) in order to study of the flight mechanics. It was equipped with fuselage, wing and empennages; it was launched with a catapult, reaching the speed of 90 *km/h*. With its help were defined the equilibrium equations which are used today to calculate the aircraft's performances.

Otto Lilienthal used for his gliders a structure from curved wooden rods inserted in sewed fabric (the lifting surface). The whole structure had a network of wires for the fabric stretching system, all being anchored by some vertical struts. The link between the wires and the wood rods was made using leather caps. It is good to be noted that the wires below the wing were from metallic wire, while those above the wing were made by thin rope. The gliders had no fuselage, but only a framing where the pilot was anchored by a harness. The whole structure was fragile, being exclusively made for gliding in calm atmosphere.

Fig. 2.2 *The Henson's flying machine with steam engine* (1842) [41]

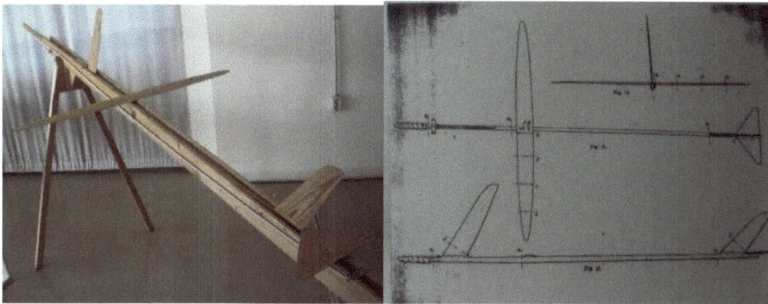

Fig. 2.3 *Aerodon* [F12]

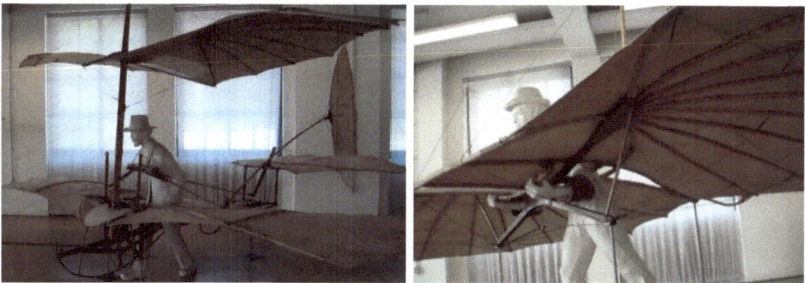

Fig. 2.4 *Two of Otto Lilienthal's gliders* [F12]

According to [67], Otto Lilienthal died in an accident caused by a strong gust of wind which leads to the wing failure. This documented one of the first human losses due to the structural failure. After [67], due to the glider fragility, at the wing failure the entire structure has lost its integrity, therefore a total failure of the early flight attempts.

Samuel Langley built a first tandem wing biplane, which immediately after its launch, crashed having an example of wing failure (Figure 2.4 illustrates the deformation of the rear wing, followed by the flying machine pitch down and crash). Pressed by his financiers, Langley hadn't enough time to do all tests for a smooth transition from scaled models to the flying machine, calculating wrongly the structure [72].

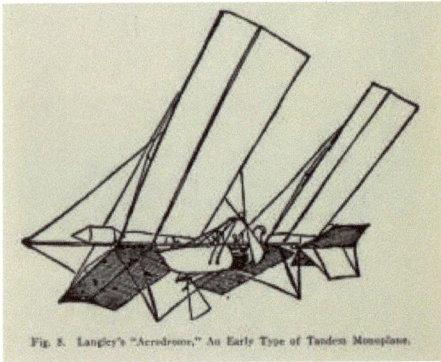

Fig. 5. Langley's "Aerodrome," An Early Type of Tandem Monoplane.

Fig. 2.5 *The biplane realized by S. Langley (1903)* [08] / [12]

According to reference [41], the attempts of Ader and Langley marked the end of a period dominated by improvised solutions and failures.

2.2. THE 1903 – 1913 PERIOD

To perform a flight, an aircraft requires a good aerodynamics, a propulsion system; a flight controls system, structural stiffness and minimum weight. Till the Wright brothers, the various flying machines realized in various proportions the above requirements. However, the Wright brothers, besides the fact that they've achieved all of them, succeeded to bring the each one to the level of acceptance to perform the flight.

Fig. 2.6 *The Wright brothers' flying machine* [67]

It can be assumed that they were able to fly not by chance. Approaching the flight in a scientific way, they were informed about the results of Otto Lilienthal studies and they benefited of the Octave Chanute's book "Progress in Flying Machines" (1894), an excellent monography, including also the author's own researches in the field of structures calculation [04], [41]. The Wright brothers continued with the aerodynamic studies of Lilienthal, they have built their own engine, they have benefited from the knowledge of the structures' theory of Chanute, and not least, they were familiar with the light structures, being bicycles builders.

Octave Chanute, railroad and bridges engineer with a successful career was the one who recommended the use of the biplane wing with longerons (spars) [12]. Each wing had two longerons of ash wood [67], interconnected by ribs. These were stiffened to planar shear stresses by placing the fabric skin with the wire oriented at 45 degrees to the direction of the spars [12]. The wings were interconnected by wooden struts connected to longerons; between them it were used as diagonals steel tensile wires in all three segments of the wing (in order to prevent the degeneration of rectangular section in parallelogram [23]). Thus, it was built a very light beam, diagonal stiffened both in horizontal and in vertical plans [12]. Basically, the wing and fuselage construction could be easily compared with the bridges construction, except that the diagonals were used only in tensile loading.

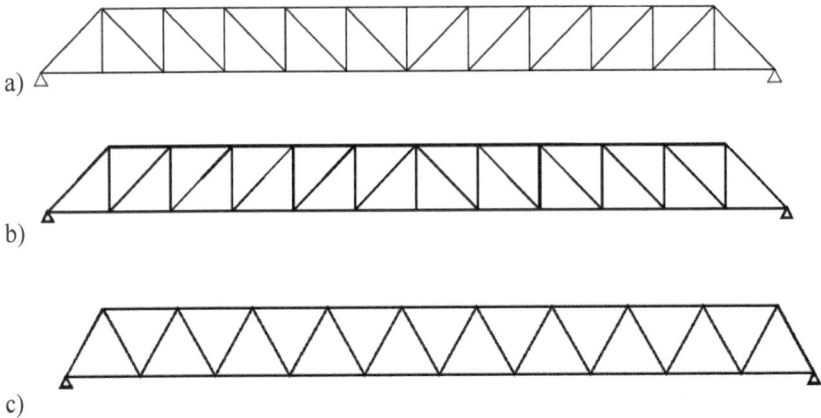

Fig. 2.7 *Common beams:* a) *Pratt – diagonals subjected to tensile;* b) *Howe – diagonals subjected to compression;* c) *Warren – without vertical columns*

Essentially, the beam formed by the biplane wing is similar with the Pratt beam (Fig. 2.7. a), containing two rows of diagonals. The use of two diagonals of wire instead one stiff diagonal is considered an innovation for the structural weight reduction [77]. This kind of beam was extensively used in the first generation of aircrafts; besides the Pratt beam, the Warren beam is also used in aviation (especially for fuselage).

"The airplane of the Wright brothers laid the basement of the future aircrafts configurations: the wing with longerons and ribs, the beam fuselage with tensile wires diagonals, the biplane with struts and tensile wires, the impregnated fabric skin, fixed twin landing gear, etc. Although, there have been variations, these components were standard until 1914" [30]. According to

Anderson [02], the Wright brothers had established a configuration which remained predominant for the next 15 years.

Although, once with the successful flight of the Wright brothers, the flight seemed to be no longer a mystery, from 1903 until approximately 1912 for the most prototypes (often called "strange machines") the major question was if they will took off or not [64].

Fig. 2.8 *Bleriot XI (1909), the first airplane which crossed the English Channel*

Bleriot VII (1907) was the first monoplane with propeller, configuration that remained at the level of idea for many years [72]. Bleriot used a system of tensile wires similar with Lilienthal, these being anchored by two struts, above and below the wing, in order to support the wing at the ground, respectively under the action of lift forces [12] (Fig. 2.8). The fuselage was constructed by ash wood with steel tensile wires. Only after 1908, the airplanes started to produce a stand-alone fuselage [02]. A much appreciated airplane was the Etrich Taube monoplane (1909), one of the first aircrafts used for military purposes. Looking like a bird, it had the wing spar from ash and bamboo ribs [72] (Fig. 2.9). The shapes of the wing and the empennage were also unique, presenting two additional vertical struts for anchoring their ribs in the console. In

Figure 2.9 it is shown a detail of bracket attaching the tensile wires to the vertical struts.

Fig. 2.9 *The wing structure of Etrich Taube airplane [F13]*

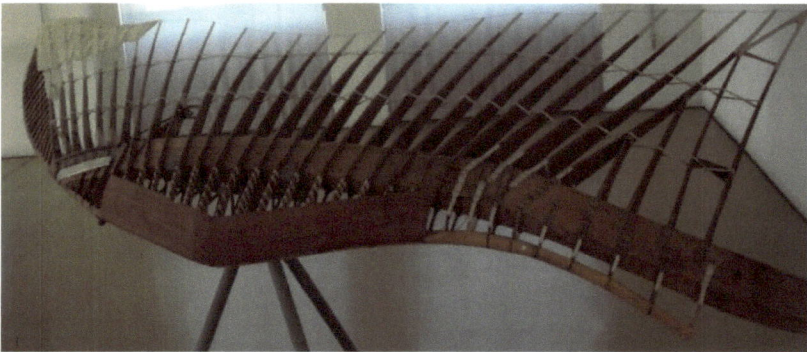

Fig. 2.10 *Wooden wing of the Geest Wolfmueller airplane (1909)* [F12]

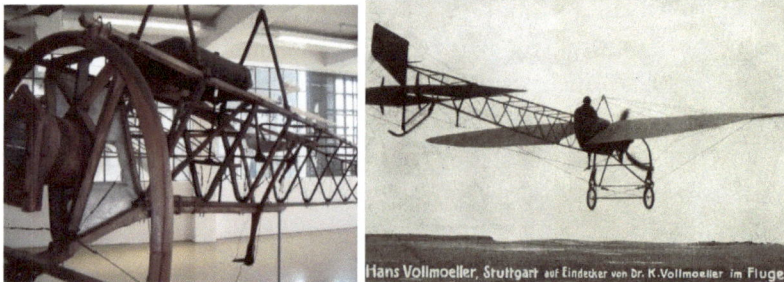

Fig. 2.11 *Vollmoeller airplane (1910) with triangular sectioned fuselage* [F12]

In 1910, Louis Breguet used the tubular spar (the wing was rotated around it, taking the role of the elevators) [28].

Another non-typical structure is met at Geest and Wolfmueller (1910) (the airplane couldn't fly, being underpowered). It distinguishes the curved wing surface, having a wooden spar box, divided ribs (nose rib and trailing edge rib) and a stabilization system which will be met to the next generation of airplanes (Fig. 2.10). A unique feature was the extreme robust wing, obtaining sufficient stiffness without any tensile wires. The heavy loaded areas were different from surfaces subjected only to aerodynamic loads showing a good understanding of lightweight structures; thus, in non-structural areas (shape ribs and wing tips) the ribs were made from rattan wood defining a ultra-light structure [Deutsches Museum, Oberschleissheim].

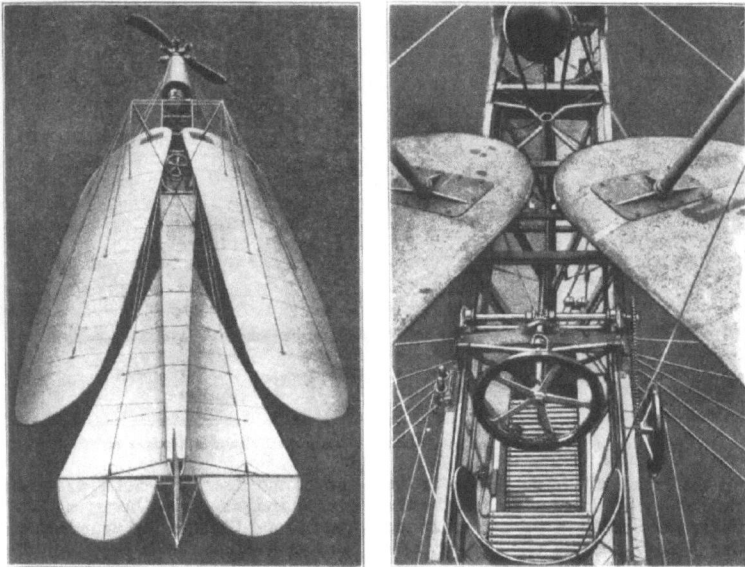

Fig. 2.12 *Airplane with retractable wings type Marcay – Mooney 1912* [F14]

The fuselage was mostly built from a square section beam. An interesting example, with triangular cross section, is the Vollmoeller airplane (1910); the fuselage frames were by curved wood ("V" shape) with diagonals steel tensile wires (Figure 2.11). Another special example is the Marcay-Mooney flying machine, with foldable wing; it was one of the first aircrafts with this ability which succeeded to fly (fig. 2.12). It should be noted that the Clement Ader plane presented this ability since 1897, but it was unable to take off.

The flying boat is a flying machine which appeared from the beginning of aviation. This kind of flying machine was introduced because of the lack of an adequate network of airports, the high mobility offered and a minimum of the facilities necessary for take-off & landing on water. The first operational flying boat was built in 1912 by Glenn Curtis, having wooden hull [35].

This chapter cannot be finish without mentioning few data about the airships. With no doubt, the count Zeppelin was the one who stimulated the development of the large aircrafts; since 1910, LZ 7 could transport 24 passengers, on a distance of 500 km [41], in a period when the airplanes were only at their beginning. Also, count Zeppelin approached the large flying machines in a multidisciplinary way, including the establishment of the commercial flight.

Zeppelin developed a series of airships with rigid frame (located inside the balloon). The large dimensions of his ships rose up to problems of local stability of the beams; the overall buckling of the structure and keeping the weight under control [67]. The latticed beam was the key element (Fig. 2.13), with similar design as for bridges, but refined for lightweight [67]. For the frame construction, it was used Aluminum alloys, the beams having sheet metal webs with holes with formed edge for weight reduction. This became a standardized weight reduction method to be used for all thin-walled structures. Progresses had been made also for the riveted joints; researches were made from the large assemblies down to the smallest components targeting the structural efficiency. For 40 years, the airships weight increased from 12 to 220 tons. The operational period was marked by many local failures, some of these leading to the entire structure failure (caused by high stress concentrators, especially at the junction of the fin with the fuselage) [67]. However, Zeppelin remains as the animator of a company which made from the air transport a standard, producing a lot of airships of various types, dimensions, payload and range, principle followed today by the great aircrafts manufacturers (Figure 2.14).

It may be concluded that in the period after the first successful flight of the Wright brothers, the aircrafts were still very different in terms of aspect, configuration and technology.

Fig. 2.13 *The structure of a Zeppelin airship* [F15]

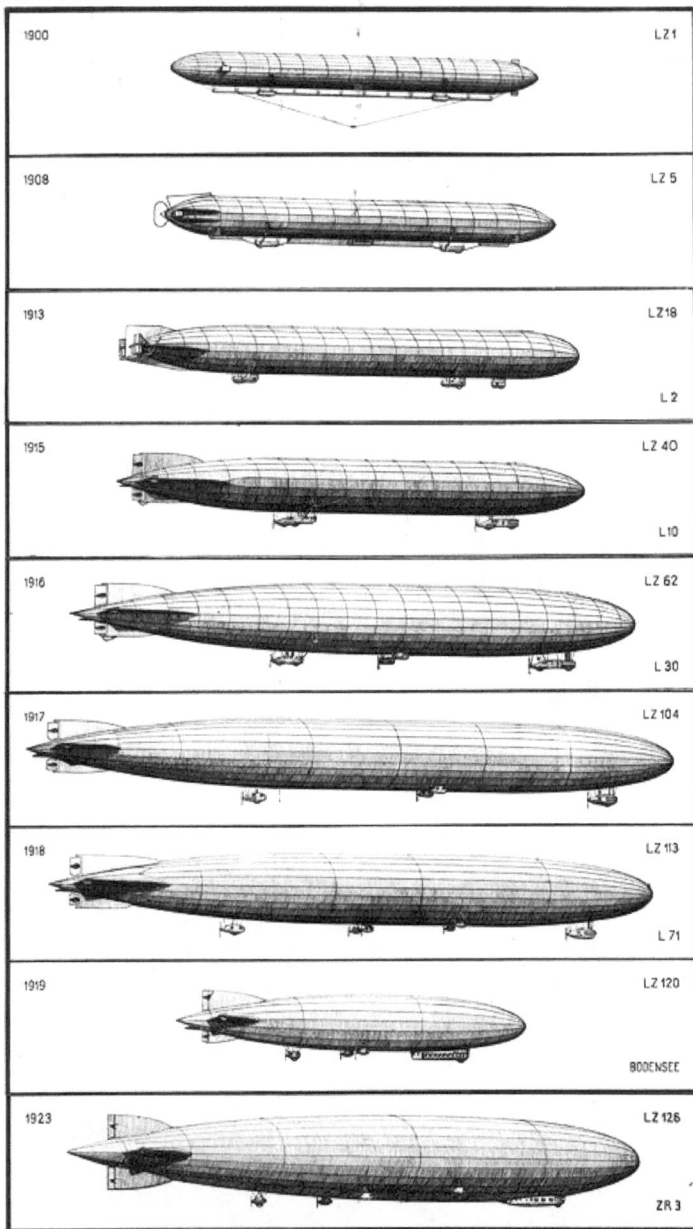

Fig. 2.14 *The range of the Zeppelin airships* [F16]

The persistence of a high number of failures indicate that, despite the excessive publicity made to the Wright brothers and their plane, the community interested in flight was not convinced of the validity of their technical solutions. These solutions set up the basement for the further development of the great aviation. As flight configurations from that time, remain the airplane and the airship, the latter being not further developed after 1937, after the tragic accident of Hindenburg, currently being used only for aerial photography and advertising purposes.

2.3. THE 1914 – 1918 PERIOD

Till 1914, there wasn't any general accepted flight configuration, flight machines being regarded more as curiosities [35]. At the end of the First World War, the aircraft was a very useful and versatile vehicle, designated to accomplish specific missions. This fact was mainly due to the manufacturing and testing of hundreds of prototypes, as a result of the massive funds of governments of the belligerent states [35].

In the First World War, the wooden planes played the fundamental role [35]. Most of the structures consisted of a wooden frame reinforced with tensile wires, covered with fabric [35] (Fig. 2.13). Some producers have used for fuselage a minimal inner structure, with skins made by plywood [30]. There were encountered also fuselages with metallic frame stiffened with wood external surface, others being entirely made of metal, with the interior of latticed beam [35].

The First World War brought the most important innovations of the aircraft: the wing in console and the frames reinforced coque fuselage [30]. The wing continued with the structure of the longerons with ribs, but the longerons were the only stressed elements [30]. The classic spar was an "I" profile, consisting in a plywood web and four wooden flanges. Another construction was that of the box spar, having two webs with the wooden flanges placed between them.

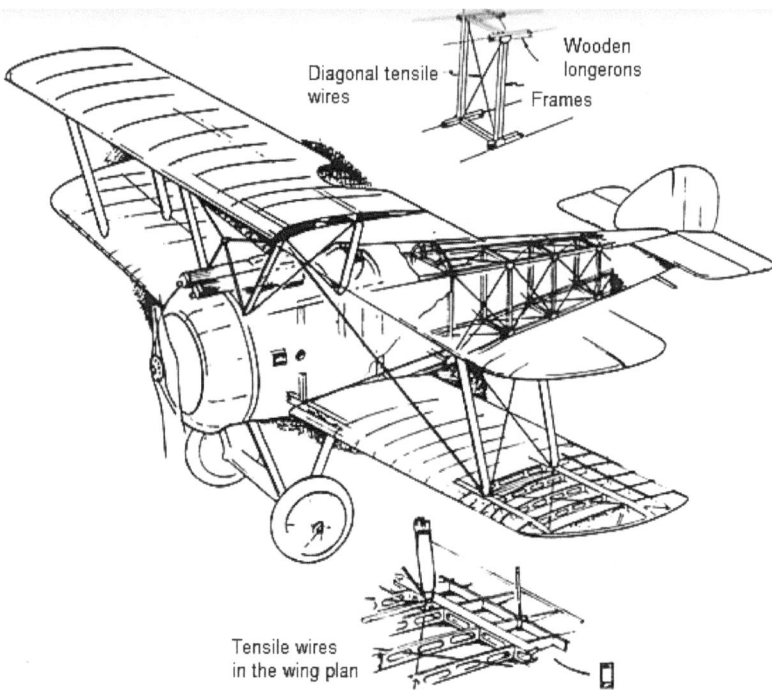

Fig. 2.13 *Typical construction of wooden biplane (Sopwith Camel, 1917)* [02]

Although, the wing in console was introduced by successful manufacturer Anthony Fokker, other producers weren't convinced by this solution, the biplane configuration remaining a standard till the World War II [35]. If firstly, the structure was covered by fabric, it was followed by the veneer skin, obtaining an increased strength; in this way, the stressed skin start to be used [30].

The request for faster and larger capacity planes led to the use of engines with more and more power. The engine thrust has the role to compensate the aerodynamic drag; therefore a better aerodynamics can eliminate the requirement for a more powerful engine. The request to reduce the power led to a more efficient configuration than the biplane, namely the wing in console, without bracings or tensile wires [28]. Junkers and Fokker demonstrated that a wing with big relative thickness does not lead to an excessive increase of the drag, allowing a stiffer internal structure.

At the beginning of the WWI the profile without thickness, with great curvature was characteristic to all aircrafts; it was believed that the relative large thick wing generates a high drag [35], [64]. Because of the airfoils with very small relative thickness (about 5%), inside the wing it was no room for an internal stressed structure (Fig. 2.14). An example is the Nieuport 17 airplane (Fig. 2.15) where, the lower wing, due to the small thickness, couldn't allow a stiff spar. For this reason, there were reported cases of structural failure of this airplane at high speed diving [35].

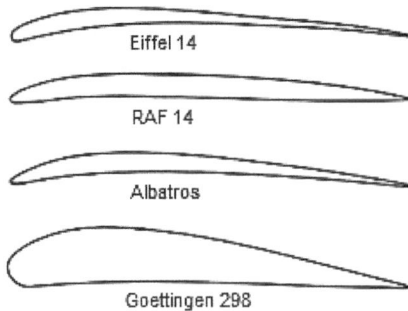

Fig. 2.14 *Examples of airfoils used during the WWI* [35]

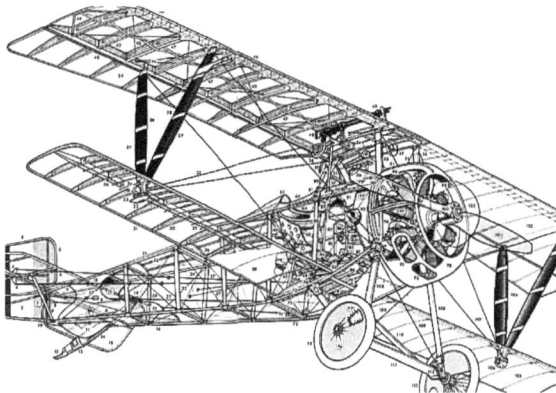

Fig. 2.15 *Nieuport 17 (1916)* [22]

At the end of the war, the Germans introduced the thick profiles, facilitating the monoplane proliferation. On the other hand, it has been proved that the profile must not be very thick in the case of metallic wing. Moreover, for the biplane, the spar loads were so low that the use of the metal would have led to an over dimensioned structure [28].

In parallel with the increasing of speed and range, a larger payload started to become a priority. In 1914, Igor Sykorsky realized Ilya Muromets, considered to be the precursor of multi-engines large aircrafts [79]. Having an integral wooden structure, this had a maximum take-off weight of 4.600 kg, a record of that period.

Between the large airplanes it distinguishes Zeppelin Staaken R VI (maximum take-off weight 11.800 kg) which had a latticed beam fuselage with square section, with lateral and inner bracings (Fig. 2.16 b). From this structure, it was attached a secondary structure of wooden frames and longerons, acting as support for the floors and skin [26].

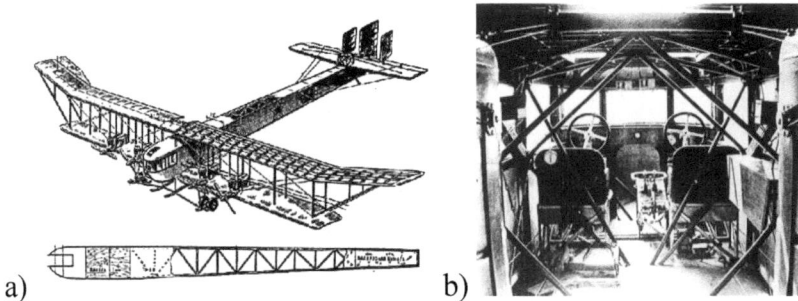

a)   b)

Fig. 2.16 a) *Four-engines Ilya Muromets airplane (1914)* [F17];
b) *The interior of the fuselage of Zeppelin Staaken R VI (1916)* [26]

A remarkable engineer of the First World War was Anthony Fokker, the Dutch manufacturer who has succeed not only to realize innovative concepts, but also highly successful aircrafts. Constructed in a number of 450 pieces, Fokker Eindecker (1915) was the first aircraft with metal latticed beam fuselage.

Using tensile wires as diagonals, it presented a unique element, being also covered with fabric [35]. The most used layout of the First World War was the fuselage covered with plywood in front side and fabric in the rear side area, up to the empennage [35].

The tri-plane Fokker Dr 1 (Fig. 2.17) introduced with great results several new features as [35], [12], [64]:

- the first wing with thick profile (Goettingen, 13%);
- the spar in two pieces, closed with plywood, forming a torsion box (caisson);
- the ribs from plywood with holes for weight reduction and stiffeners for shear loads;
- the leading edge made by plywood;
- the fuselage, aileron and empennage from welded tubes frame;
- the first structure without tensile wires;
- the number of struts lowered to the minimum, employing a single pair of struts at the ends of the wing.

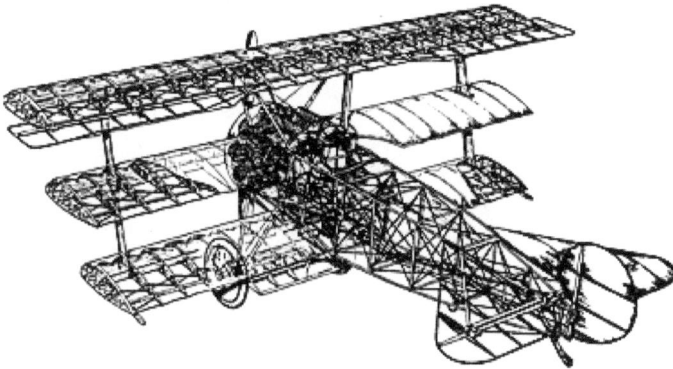

Fig. 2.17 *Fokker Dr-1 (1917) – with tubular pillars* [12]

The drag of the tensile wires represented a disadvantage, because of they were needed in a big number (a number of 26 tensile wires of 6.25 mm diameter, is equivalent with the drag of an empennage section of 600 mm chord and 12% thick [35]). Therefore, the elimination of the tensile wires was a real advantage. These could be replaced only by a stiff wing structure and thereby a higher relative thickness of the airfoil [12]. These solutions started to multiply at the end of World War I. It should be noted that the Sopwith Camel airplane (1917) introduced the tensile wires with aerodynamic profile (with symmetrical airfoil), with the drag of 10 times smaller than the cylindrical ones [35], [64].

A unique feature of Fokker Dr 1 was the composite wing spar, with a hollow structure, showing a high effective bending and torsion behavior (Fig. 2.18, Fig. 2.19.). The common spar was replaced by a system of double box spars, jointed into a robust torsion box [02].

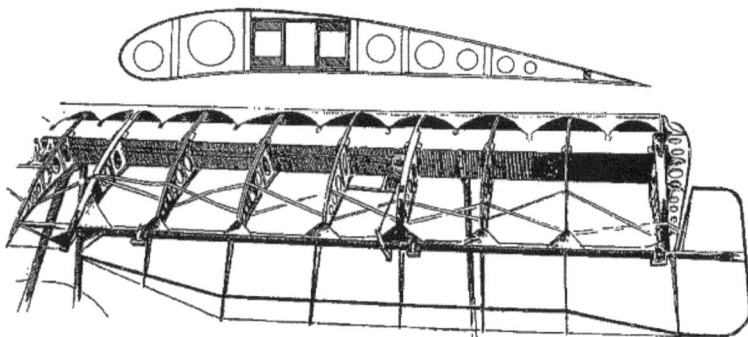

Fig. 2.18 *Detail of the wing of Fokker Dr 1 airplane* [12]

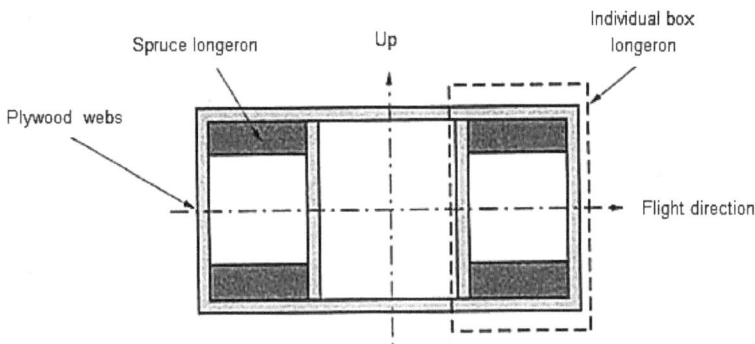

Fig. 2.19 *The box spar of the Fokker Dr 1 airplane* [35]

Fokker D VII (1918) was a monoplane, without tensile wires or struts, using a single wing in console. It was the most advanced aircraft of that time, although, it had structural problems due to the aero-elastic phenomena [12]. The D VIII model introduced the trapezoidal plan wing, with a sloped latticed beam longeron (the height was decreasing towards the wing tip). Thus, Fokker obtained the wing weight reduction and the moment reduction in joining area [35].

Fokker D VIII successfully employed the wing in console, with thick profile and plywood skin. Although, the D VIII model came too late to influence the end of the war, Fokker developed his ideas within some passenger airplanes which marked that age [30].

Fig. 2.20 *Anthony Fokker in front of D VIII airplane, proving the robustness of the wing in console* [30]

In 1916, Junkers built the J1 aircraft (fig. 2.21), having the first integral metal structure [35], [30]. Although only the prototype was manufactured, this encompassed a number of key features as [35]:

- monoplane with internal structure (without tensile wires);
- thick airfoil wing, variable in size (17% at fuselage junction and 12% at wing tip);
- the first airplane with stressed skin [30].

The skin was made of steel sheet, because Duralumin didn't have yet the technology to achieve sheets raw material (presenting delamination). The steel skin of 0.1 ÷ 0.5 mm (extreme thin) was doubled by an inner structure of welded corrugated sheet metal [28], [2]. Having an inner structure of steel tubes, the J1 model was slow and hard to maneuver, presenting many maintenance problems due to the welds quality [30].

Fig. 2.21 *Junkers J1, with integrally metal construction* [78]

Although, it had a large relative weight, the wing had a high durability, the same wing being used by Ford Trimotor (1926) [35].

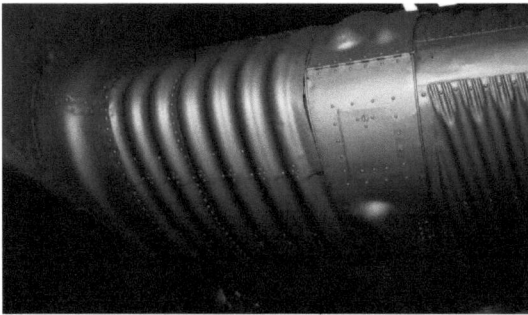

Fig. 2.22 *The leading edge from corrugated sheet metal (Junkers)* [F18]

Junkers was a visionary engineer, only in the 1930's the wooden aircraft starting to be replaced by metallic ones [02].

Due to problems with J1 model, Junkers definitively abandoned the steel, immediately adopting the Duralumin aircraft [02]. The following aircrafts had the skin made of corrugated sheet metal (Fig. 2.22), this becoming emblematic for the whole family of Junkers airplanes. The corrugated sheet metal skin increased the longitudinal stiffness, but the overall weight increased with 20 ÷

40%, too [35]. In reference [47], is mentioned that as a result of experience with J1 model, Junkers adopted the wings able to take over more efficient the bending moment by the meaning of a tubes system.

Fig. 2.23 *Junkers J 2 – The welded central section* [78]

Figure 2.23 shows the central section of the Junkers J2 airplane, consisting of a welded frame where the wings, engine and the rear fuselage are connected. During the time, the wings junction was made either on fuselage with dedicated fittings, or positioning it under or above the fuselage, respectively. The idea of a central section (Central Carrythrough Structure) was reconsidered in time, now being used on all modern passenger aircraft with low wing. Actually, the wing mounting is no longer realized in concentrated points, but on the whole wing section contour. In Figure 2.24 is presented the central section of the Airbus A380 airplane, which has also the role of fuel tank.

After Junkers, Dornier and Zeppelin build the Dornier D. I., an airplane with metallic wing, except the control surfaces which were covered with fabric (Fig. 2.25). The fuselage was fully metallic, with a stressed skin, and integrated fin [33].

Fig. 2.24 *Airbus A380 – The central section* [78]

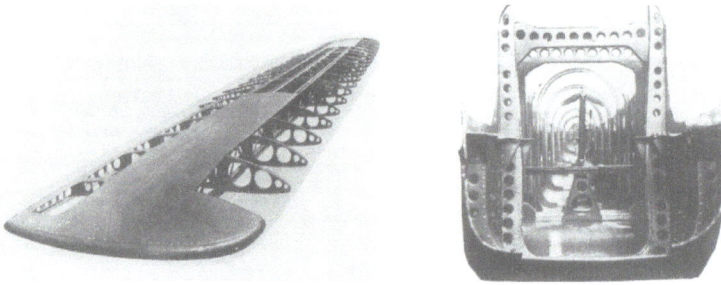

Fig. 2.25 *The wing and the fuselage of Dornier D.I (1918)* [73]

The first use of frames reinforced coque construction belonged to the Swiss naval engineer E. Ruchonnet, in 1911, with the Cigare airplane, by applying an orthogonal system of veneer layers on a circular wooden frame [05].

In 1912, Louis Bechereau was able to overtake the existing aircrafts, with a coque fuselage from plywood, reinforced with bulkheads and longitudinal stringers [30], [35]. The Deperdussin racing aircraft had the fuselage skin made by 3 bonded layers of veneer of 1.6 mm, formed using an outer mold [28]. The result was a very stiff structure, with a smooth aerodynamic shape. The two Deperdussin models built in 1912 were remarked by breaking several speed records [28], [30], but in spite of this, there were further serial production.

The first airplanes with monocoque structure were built during the World War I, only in small numbers, in France and

Germany [28]. Morane-Saulnier N (1914), also called Morane Monocoque (concept inspired from Deperdussin), had the monocoque fuselage with steel rings and metallic tensile wires [28]. The very poor maneuverability made that this aircraft to be quickly forgotten [22], and with it, the monocoque concept.

Fig. 2.26 *Albatros DV (1916) wooden monocoque fuselage* [22]

The German Albatros airplanes recorded a great success (1916) (Fig. 2.26). These had monocoque fuselage, built from bonded plywood straps on a wooden structure of frames and stringers [02], [30], [35].

An interesting solution was met to Jimmy Pfaltz model (1917) – the longerons attached to circular frames formed a beam; this was reinforced with two thin layers of bonded plywood, the external layer being from fabric (Fig. 2.27). The result was an aerodynamic shaped fuselage, very uncommon for that period. Over time, the fuselage was presenting deformations due to improper treatment of the wood and to the water absorption [02].

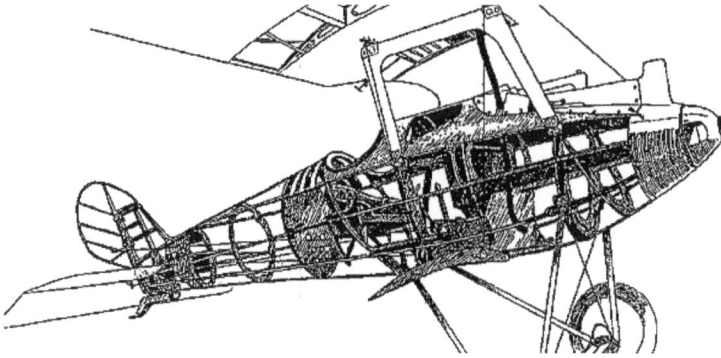

Fig. 2.27 *The fuselage of the German airplane Pfaltz D-III* [02]

In 1918, Lockheed has patented the frames reinforced coque structure, made by two halves, formed in reinforced concrete molds by pressing with a rubber quilt. This allowed the fabrication of constant thickness self-stiffened skins, requiring a reduced internal structure; in this way it was obtained a smaller cross section than the similar aircrafts, consequently a lightweight structure [30].

Although, the monocoque and wing in console advantages were obvious, the production of biplanes with latticed beam structure continued at a big scale [30]. A possible explanation is the huge market request, leading to the less interest of producers for concepts difficult to be industrialized, even they proved their validity.

Dornier and Junkers were the promoters of the metallic structures. Due to the lack of some alloys allowing the manufacturing of structure competitive with wooden airframes, the stressed skin fuselage appeared later [35]. It is good to mention that the sheet metal was used only as a thermal protection of the fuselage and the pilot against the exhausted gases and engine high temperatures (steel, chromium and nickel).

The war has left as the consecrated solution the biplane with bracings and tensile wires, with fabric covered wooden structure as the best compromise between structure strength, the minimum weight and the aerodynamic qualities [35]. This proved to be the most secure configuration, being considered a structure that could not fail [12]. As a result, the biplane has been used for many years after the end of World War I [35].

2.4. THE 1919 – 1929 PERIOD

After the World War I, it followed a weak period for the production and development of aircrafts. The mail and passengers transports were expanded, many airways openings being recorded. The development was stimulated by the premiere flights and air racings funded by US Government or by the air armies (France, England and Italy) [35].

In USA it were elaborated the first regulations for the aircrafts purchased by army; there also appeared the first regulations for the aircrafts airworthiness and the licensing of the pilots. The regulations have stimulated the settlement of the general aviation. Creating the legislative frame, the safety of flight increased, together with the people's confidence in aircrafts. All the aspects above stimulated the development of aviation [35].

In 1919, Short's produced the Silver Streack airplane, having the wing longerons from steel tubes, with Duralumin formed ribs and skin, riveted together [47]. The formed ribs were to become a standard, being met in all modern metal aircrafts. Since 1920, it was expected major improvements of the metallic construction, based on the prevision of the wooden structures end [64]. However, because of the technological limitations and the lack of appropriate materials, the wooden airplanes have continued to be produced in a large number.

In the 1920s, many successful monocoque aircrafts were built in France (Bleriot, Bernard, Hd780, Nieuport) and UK (Curtiss and Short) [28]. The aircrafts with an aerodynamic plywood surface from Germany have demonstrated the benefits of the low drag [28], which has stimulated the producers to cease covering the aircrafts with fabric.

The non-stop flight New York - Paris from 1927 (Charles Lindberg/Ryan) had a great media impact, thereby increasing the public trust in this new transportation way [35], [41]. Ryan made a great advertising to the monoplanes; the biplanes starting enter in a shadow cone [35].

Duralumin was not extensively used due to the lack of suitable corrosion protection; this was realized only in 1923 in England (anodization) and in 1927, respectively, in America (aluminum plating). The Boeing B247 model (1933) was the first civilian airliner which used metal sheets plated with Aluminum (CLAD) [02].

a)
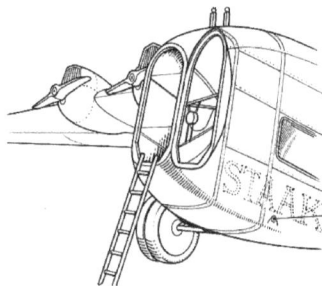
b)

Fig. 2.28
a) *Zeppelin Staaken E4/20 (1919)* [78]; b) *detail with access door* [02]

In 1919, the German Adolf Rohrbach designed the Zeppelin Staarken E4/20, a passenger airplane by Duralumin with the thick wing and the fuselage with monocoque, whereof the prestigious magazine Flight said in 1921 that "this monoplane represents an important progress for what the airplane represents in our days" (Fig. 2.28 a, b). To Rohrbach is attributed the wing concept with stressed skin [47]. The wing had box structure (common to both left and right wing) and stressed skin stiffened with leading edge and trailing edge ribs [60]. The fuselage had an advanced aerodynamic shape. In order to maintain the continuous monocoque, the access was made through the aircraft nose (the large openings of the passenger or cargo doors lead to the significant added weight because of cutouts margins stiffening). This concept will be later used in all the major cargo transport aircrafts (C5 Galaxy, Beluga, Airbus A400M - by tail etc.). Due to restrictions of the Treaty of Versailles, the prototype was destroyed in 1922 and, with this, an airplane that was "tens of years before the passenger aircrafts of that time" [60]. Moreover, the aviation historians assert that "Rohrbach could become for Germany what Boeing and Douglas represented for USA" [60]. In 1921, Fokker builds F2, an airplane designed to transport 4 passengers, monoplane, wing in console, without struts or tensile wires; the wing was from wood, and the fuselage had metallic structure covered with fabric [35].

In 1922, the Latecoere Lat 6 aircraft had the fuselage by Duralumin covered with steel sheet [28].

After Claude Dornier practiced the Duralumin constructions working for Zeppelin, he designed an integral metallic flying boat before the end of World War I, the first flying machine being produced only in 1922 (Dornier Wal). It had a configuration that anticipated the future design of passenger flying boats [60].

In order to increase the speed, Boeing introduced on the 40B airplane (1925) a wooden wing (Fig. 2.29). Because of the increased drag, the stiffening in the horizontal plan was no longer possible only by the fabric skin, needing additional diagonal tensile wires between the longerons (the concept of beam stiffened against drag - drag truss) [12].

The Nieuport NiD 52 airplane (1927) replaced the wooden wing and fuselage with the fabric covered metal construction.

The monoplane problem was the wing in console, a matter that was solved in early 1930s [12]. One of the main innovations was the stressed skin, able to withstand to the planar shear stresses [12]. The fabric has been replaced by plywood or sheet metal, providing the buckling strength for an additional weight.

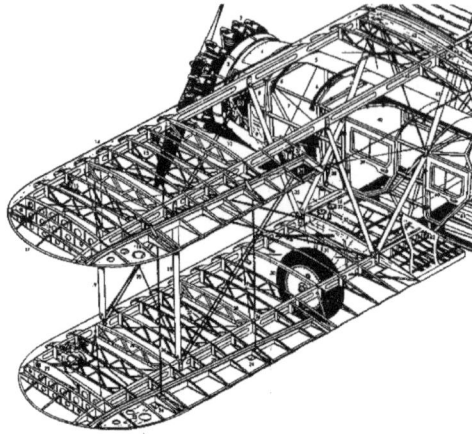

Fig. 2.29 *The wing of the Boeing aircraft model 40* (1925) [22]

An example is the Lockheed Vega 1 (1927) which had the wing with plywood skin, together with spruce longerons forming a stiff torsion box, with good behavior both to bending and torsion loads [12]. A specific feature was the fabric applied on the plywood skin panels (Fig. 2.30). The fuselage with wooden monocoque structure, realized by two halves (for the productivity purpose) had the many characteristics of the modern passenger aircrafts [28]. As per reference [47], the monocoque structure with skin of 2.5 mm thickness was so stiff that the frames size was reduced.

In that period, the flying boats were among the heaviest aircrafts, for this reason, it may be stated that, together with the bombers, these have stimulated the development of large aircrafts. In fact, in those times, the transoceanic flight was unthinkable without a safety solution (allowing the sea landing) and without a payload to justify the fuel consumption by commercially point of view. Curtiss NC 4 was the first aircraft that crossed the Atlantic, in 1919. It is good to be noted that the flight, in addition to the pilots, a radio operator and navigator included two engineers responsible with the in-flight malfunctions fixing. This aspect is relevant for the safety of flight of those times.

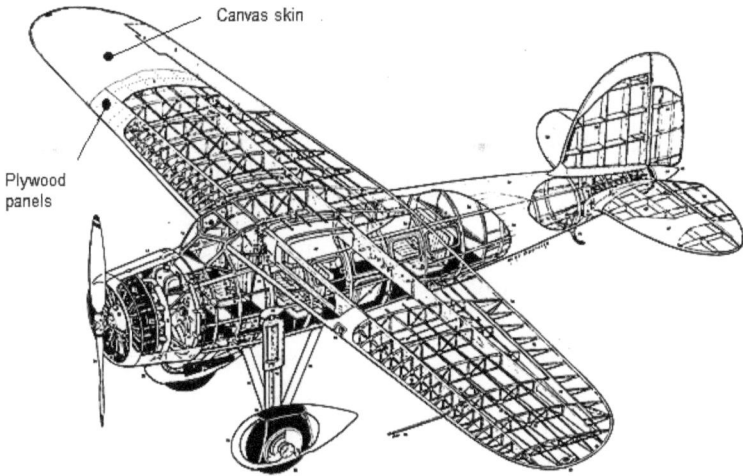

Fig. 2.30 *The Lockheed Vega aircraft (1927)* [12]

The first flying boats were biplanes with wooden structure. A disadvantage of those was the water absorption, which over the time represented an unwanted weight [35]. Another problem was the sea water, which is very corrosive. Before the use of Duralumin as stressed skin for the hull, wooden skin plated with Aluminum sheet was used.

Fig. 2.31 a) *The Loening 0A-1C amphibious aircraft (1924);* b) *The Sikorsky S-38 amphibious (1928)* [35]

In 1924, the Loening 0A-1 C amphibious airplane appears (with retractable landing gear), with wooden hull structure covered with Aluminum skin (Fig. 2.31 a) [35]. The wing had wooden longerons, Aluminum formed ribs (a technology which prefigured the standard manufacturing of the Post World War II period) and fabric skin.

In 1925 in the US, the first PN-9 flying boat with full metallic hull was built. An interesting configuration it observes at Sikorsky S38 (Fig. 2.31. b), one of the few amphibious passenger aircrafts, with wooden fuselage covered with Duralumin and the wing by steel studded structure [35].

In 1928, the Martin PM-1 flying boat had a fabric covered wing with steel structure, with a small number of tensile wires and the stiffening pillars [35]. In the period 1934 – 1938, the flying boats were improved by the removal of side floaters (by replacing with floating sponsons), the full metallic structure and the wing in console attached to the upper side of the fuselage [35].

In 1926, Fort Trimotor produced an airliner with full metallic structure, having the skin from corrugated sheet metal (the transverse buckling behavior was very good, having also additional weight).

An interesting configuration is met in reference [29], the fuselage presenting a lower beam (similar to gliders); the fuselage body was made from a system of frames, light stringers and external skin (Fig. 2.32). There is no information available about the aircraft type or its producer (probably from Russia).

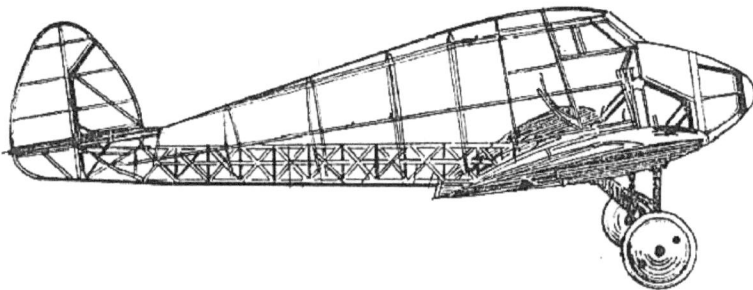

Fig. 2.32 *Airplane with beam stiffened fuselage* [29]

THE EVOLUTION OF THE AEROSTRUCTURES. CONCEPT AND TECHNOLOGIES

2.5. THE 1930 – 1938 PERIOD

After 1930, the biplanes weren't largescale used; in the same year the metal starts to be introduced as material for the aircraft structure [23], many companies starting to develop airplanes with full Aluminum structure [12]. Firstly, there were replaced only the wooden elements such as longerons and pillars, by sheet metal formed parts [23].

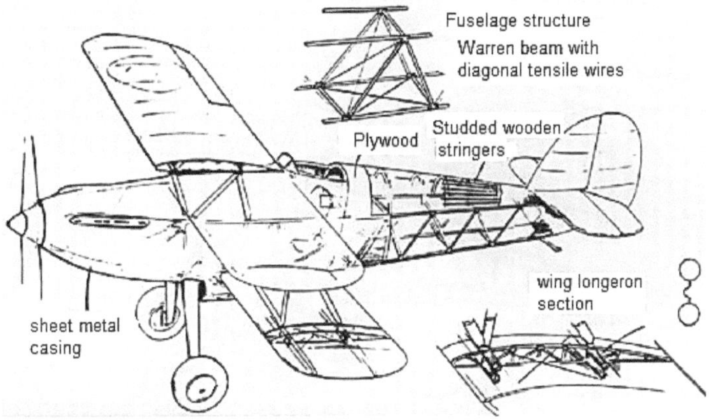

Fig. 2.33 *The Hawker Fury (1931) – fabric covered metallic structure* [13]

Hawker Furry (1931) had a Warren latticed beam structure (members attached with bolted clamps). The diagonals of the frames beam were from tensile wires. The fabric skin was installed on a system of wooden stringers, attached from the fuselage beam. In the engine area, the fuselage was covered with sheet metal cowlings without a structural role (Fig. 2.33).

After 1930, in U.S.A, the monoplanes appeared as a result of U.S. Navy requirement to increase the performances, but also as a result of the commercial pressure. In 1931, Consolidated Commodore 32 was built, a flying boat with full metal hull and wing with metallic structure covered with fabric, except the leading edge which was metallic [35]. Starting with 1930's, Douglas Dolphin and Grumman G21 had the entire wooden monoplane wing in console, and the entire hull by Aluminum [35].

83

The concept of stressed skin was introduced by Junkers in 1923: "Theoretically, it seems that the best concept is the system of supporting coverings; in this way, all bending, shearing and compression loads are taken by the skin" [02]. It is worth to note that Junkers used only occasionally this concept [02].

The term of stressed skin was introduced by Rohrbach, in 1924, being used for the flying boats [02]. His design of wing spar together with the panels of stiffened skin on four sides is considered as revolutionary [02].

Until the 1930's, it was considered that the internal structure of the wing together with skin were sufficiently stiff to prevent the buckling – in the conditions where the local buckling was considered as a structural failure [02], [28]. In 1928, Herbert Wagner demonstrates that a beam with perpendicular elements and a thin sheet metal web (with stabilized edges) doesn't fail as a result of the local buckling, but, moreover, it become stiffer [02], [28]. It has been demonstrated that the stability loss leads to the supporting loads up to 100 times larger than those that generated the instability, the structure remaining in the elastic domain [28].

In 1930, Schuman and Back have experimentally demonstrated that a thin wall plate laterally supported subjected to the buckling, reaches the maximum stiffness after the stability loss, and this limit can be exceeded in certain cases [28].
When the skin is able to support large forces, the distance between the stiffeners may increase, resulting lighter structures [02].
By accepting that the buckling of the web spar is not a negative phenomenon, the Wagner beam revolutionized the airplanes structure in the 1930's, being very used nowadays [02].

The first airplanes with monocoque structure were constructed by wood due to the reduced manufacturing costs (especially for tooling) and high stability for the same weight [28]. A cylinder with the radius of 900 mm by spruce plywood of 4.8 mm has a critical buckling force of 24 tons force. An Aluminum cylinder of 0.8 mm (of equal weight) has the critical buckling force less with 33% [28].

The wood has a low buckling strength relative to the compression strength, for that reason, the wooden monocoque fuselage skin was thick, therefore heavy; the wood wasn't used in an advantageous way [28]. The Aluminum fuselage could use effectively the material only by adding the longitudinal stringers to improve the stability [28].

In the case of the monocoque fuselage, the area between two stringers behaves similar with a Wagner beam (Fig. 2.34).

Fig.2.34. *Static loading of the Heinkel He -111 fuselage* (1936) [28]

Since the 1930's, efforts were made to reduce the skin thickness, by increasing the stringers section, a process justified by the poor buckling behavior of the metal sheet panels [28]. "It has been demonstrated that thin metal sheets do not constitute an impediment and that after the local buckling occurrence they withstand to increased loads. This was at the expense of the aerodynamic surfaces. To delay the local buckling occurrence, it was reduced the distance between stringers. In order to maintain a minimum weight, the stringers section was reduced. In order to prevent the overall buckling of the assembly formed by the stiffened skin with stringers, the distance between frames was reduced, followed by the frames section decrease. It was obtained a monocoque structure, with more homogenous structure, with more stiffening elements and a reduced and a lighter section [28].

a) *The general buckling* [28]   b) *Local and general buckling* [44]

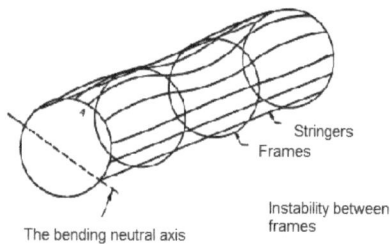

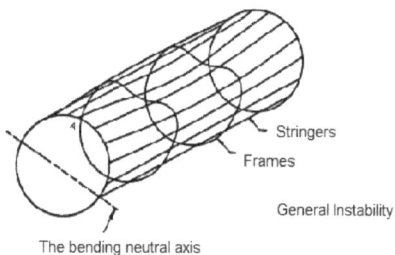

Fig. 2.35.

For a slender structure, the general buckling phenomenon may appear. It involves both the skin and the stiffeners (stringers and circular frames) (Fig. 2.35 a – the frames are not visible because these are located on the external side). In practical design, the buckling between frames is accepted, but, the general buckling is forbidden (Fig. 2.35 b).

In order to prevent the stability loss, the large openings in the fuselage are stiffened by adding longerons, and stiffened frames, respectively (especially the cargo doors and the bombs bays).

For the wing, the monocoque structure is similar to the fuselage, using the corrugated sheet metal or stringers placed along the span. The longerons were manufactured either with thin webs to take over the planar shear forces (shear webs) having a part of stressed skin, either with two or three stiff longerons, the skin taking over only the planar shear forces [28].

Between 1917 and 1933, Junkers built aircrafts with corrugated Aluminum sheet with tubular longerons (latticed beam) for wing, and omega profiles for fuselage (Fig. 2.36, 2.37). Although, it was plated with sheet metal, the Junkers fuselage cannot be considered a monocoque structure.

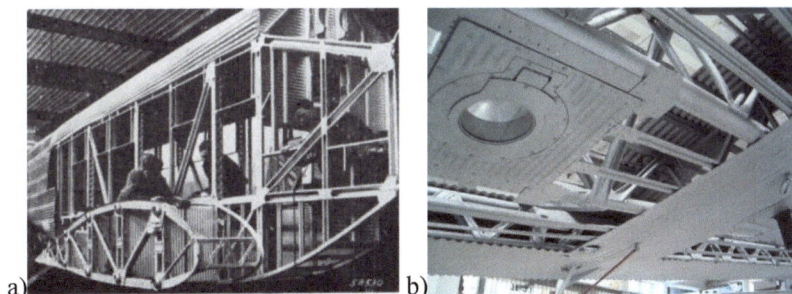

Fig. 2.36 *Junkers Ju-52 (1930) with corrugated sheet metal;*
a) *The fuselage with omega Aluminum longerons* [28]; b) *Wing detail* [F18]

Fig. 2.37 *Typical wing structure of Junkers aircrafts* [20]

Although, the monocoque was already practiced, due to the complicated system of stiffening against buckling, the big weight, the manufacturing costs and long period of development, the producers were determined to search other alternatives. An example is the wing of J. Stieger (1928) which consisted in a mono-longeron and a diagonal tensile wires system between the ribs. Over this structure, a fabric surface was tightened, having also a stiffening role (Fig. 2.38). In this way, the wing became lighter and it had a better aero elastic behavior than other existing solutions.

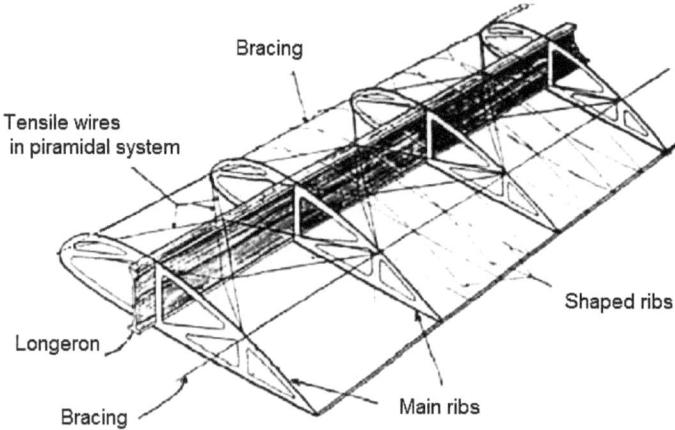

Fig. 2.38 *The Stieger mono - longeron wing (1328)* [21]

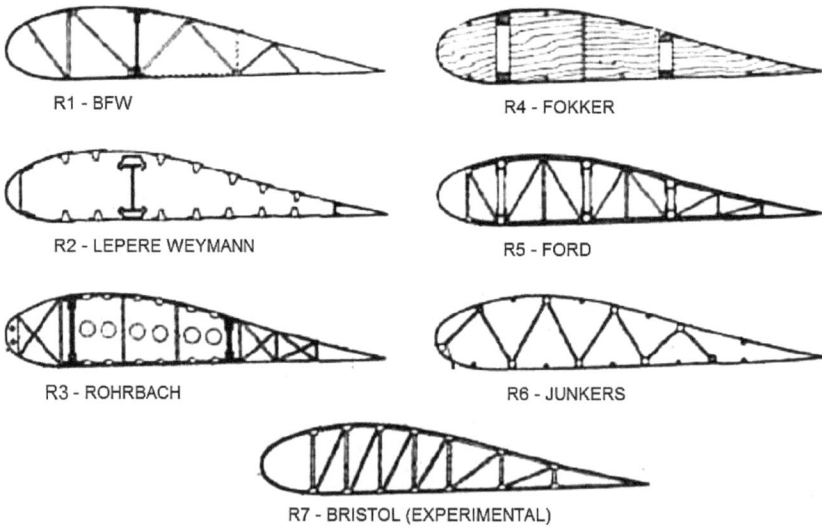

Fig. 2.39 *The constructive solutions for wing with stiffened skin* (F21)

In 1932, the Flight magazine made a synthesis of the configurations used for the wings produced by various manufacturers (Table 2.1 and 2.2). In Figure 2.39, the constructive solutions with stiffening skin are presented. It is observed that most of the concepts present a rib of latticed beam type. The rib has the role to provide the airfoil contour and to transmit the efforts from skin to longerons. Thus, the construction of a rib should be closer to the one of the diaphragm and surely not one of a beam. The solution of beam rib was used in for the wings without spar - like Junkers, where the stiffening along the wing span was provided by tubular stringers.

Table 2.1   The synthesis – wings with stiffened skin [Flight, 1932]

No.	Producer	Material	Particularities	Remarks
R1	BFW	Duralumin	Mono longeron, Stressed leading edge	The skin buckling requires a second spar
R2	Lepere Weimann	Duralumin	Mono spar, Stringers skin	The big number of stringers leads to the weight increase, Poor torsion behavior
R3	Rohrbach	Duralumin	Bi – spar, Wing box	Weight increase due to the stringers for the skin stabilization
R4	Fokker	Wood	Thick skin	Stringers are not required, The lightest design
R5	Ford	Duralumin	Tri-spar of Warren type	Difficulties of the longerons calculation
R6	Junkers	Duralumin	Pyramidal structure, Longerons flanges by tube	Complex structure, Heavy skin
R7	Bristol	Duralumin	Multi-spar (Similar to Ford)	The airfoil is changed by wing buckling, The airfoil can be kept only by using stiffer ribs

Among the presented solutions, the one proposed by Weymann Lepere (R2) and the central structure of the Rohrbach wing were closer to the modern aircraft wing. The Fokker (R4) wing has a rib made of wooden plate, composed by three segments – leading edge, central and trailing edge. The concept is a very actual one within the wing with composite structure, the rib being a sandwich structure with the external plies from carbon fiber, with a foam core.

The latticed beam is no longer used in the actual aircraft wing structure. However, the large structures are no longer built only using the monocoque concept, in the much stressed areas where used bracings, which are closer to the latticed beams. The current airplanes use solutions generated by the topological optimization software, the resulted structures being similar to the latticed beams. As manufacturing technology, the new beam ribs are made from a single part by machining, using the riveting only for the assembly. The wing with fabric covered skin requires a stiffer structure versus to the one with stiff skin, but the skin was lighter (Fig. 2.40).

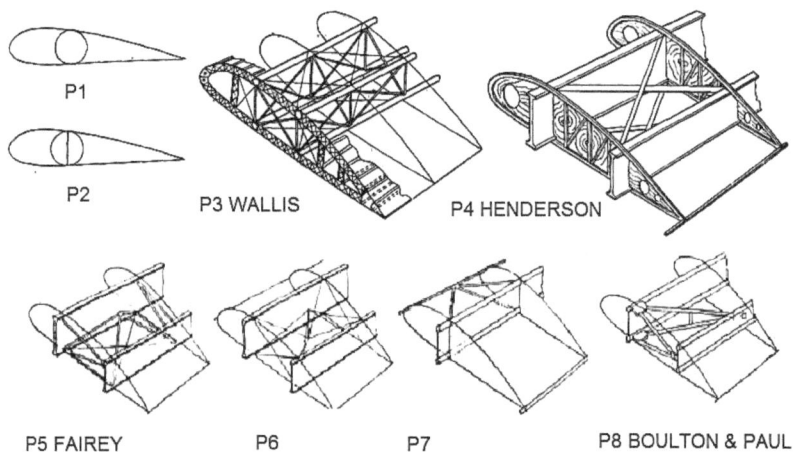

Fig. 2.40 *The constructive solutions for the fabric covered wing* (F21)

Table 2.2   Synthesis of the fabric covered wings [Flight, 1932]

No.	Producer	Particularities	Remarks
P1		Tubular spar	Requires many ribs Poor bending behavior
P2	Duncanson	P1 + *I* shaped spar	The dimensioning for the torsional strength will lead to large bending deformations
P3	Wallis	Longerons with tubular flanges with tensile wires	At bending loading, the upper skin becomes soft and it cannot take over the torsion
P4	Henderson	Bi-spar with ribs in three segments	Poor torsion behavior
P5	Fairey	Bi-spar with bracings and tensile wires in a pyramidal system	Due to the asymmetry, the drag will produce the wing torsion
P6		Bi-spar with tensile wires in pyramidal system	Stieger wing versions
P7		Mono-spar with bracing of leading edge and bracings in a pyramidal system	Stieger wing versions
P8	Boulton & Paul	Bi-spar with bracings that forms a Pratt beam in the wing plan	Poor torsion behavior

There are a large number of constructive solutions, which today seem complicated and risky (especially those with tensile wires). It is worth to note that in the 1930's, there were still supporters of the mono-spar type wing with fabric skin and an internal network by tensile wires. It can be stated that the way to monocoque was not a straight one, the concept succeeded to be generally accepted only after it demonstrated undisputed performances, an improved and cost savings manufacturing methods and a profitable commercial operation.

After John K. Northrop realized the first fully metallic flying wing, in 1929 [28], he built the Alpha airplane (1931) with monocoque structure with stressed skin [35]. The wing had no spars, but it had vertical walls which took over the shearing; the skin stability was provided by stringers made of formed sheet metal [28] (Fig. 2.41); on the upper side, these were in a larger number because of buckling sensitivity. The stressed skin withstands the bending, tensile and compression loads [47]. The fuselage was joined using flush head rivets and the wing ribs had formed flanges to be riveted to the skin [47].

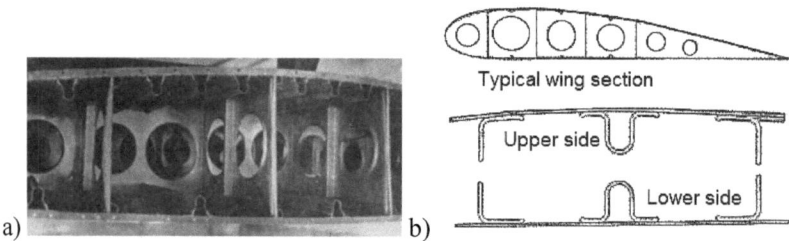

Fig. 2.41 a) *The inside of the Northrop Alpha wing (1930)* [F26]; b) *Wing section and detail [F27]*

Although it was only designed for the mail transport or a maximum of six passengers and it was produced only in a total of 17 units, Northrop Alpha brought together many innovative features which were later found on the large aircrafts. Together with the wing configuration, the central carry through structure integrating the lower fuselage panel and the wing root was a solution that offered a particular strength for a lightweight design (Fig. 2.42).

In 1932, the Hall P2H-1 flying boat was built having a full metallic hull and wing, being made in a single piece.

Boeing P26 (1932) was the first fighter aircraft with low wing and metallic monocoque structure; it appeared in a period when the producers had a different vision [22]. P26 (Figure 2.43) was a contradictory aircraft, having advanced structural elements, but also a multitude of external tension wires for wing, remaining a transition concept to a new generation of fighter aircrafts [22].

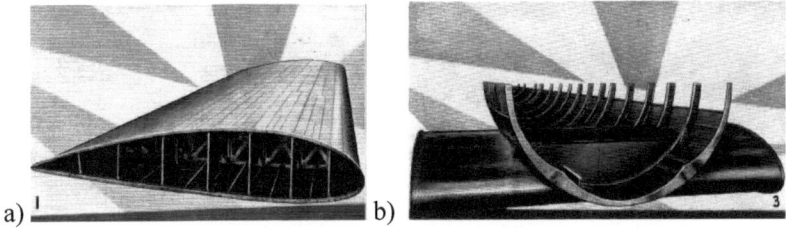

Fig. 2.42 *Northrop Alpha (1930):*
a) *Inner structure of the wing;* b) *The central section [F22]*

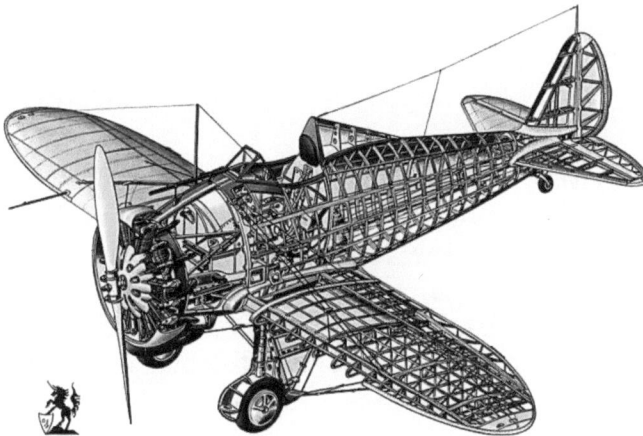

Fig. 2.43 *Boeing P26 (1932)* [F23]

Since 1936, in the U.S. all fighters were built with aluminum monocoque [35]. The airplanes with latticed beam fuselage continued to appear (Fiat CR42 Falco and Polikarpov I-153 biplanes in 1938 and Yakovlev Yak-3 in 1943).

In the case of the well-known Romanian IAR 80, the front fuselage (before the pilot's cabin) was of latticed beam type, the rear side being of monocoque type; IAR 80 could be considered a transition airplane between the two types of structure. Another example of the coexistence of two concepts is shown in Figure 2.44, combining the fuselage monocoque with a latticed beam structure for the empennage.

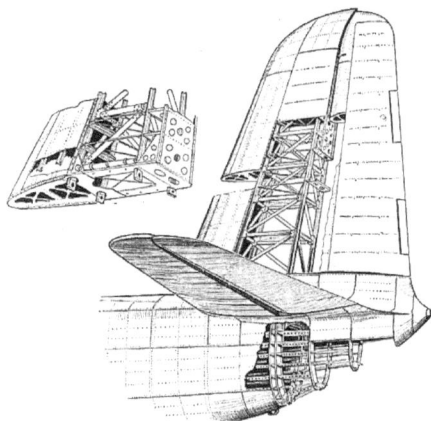

Fig. 2.44 *The Shorts Empire Boat empennage* [F24]

In 1933, the passenger aircraft Boeing 247 (fig. 2.45) combined almost all modern elements of that time [35]. From the structural point of view, even B247 was fully metallic, its construction was similar to the wooden airplanes developed by Fokker or Lockheed, having the spar as main stressed element, the skin having only the role of additional stiffening [30].

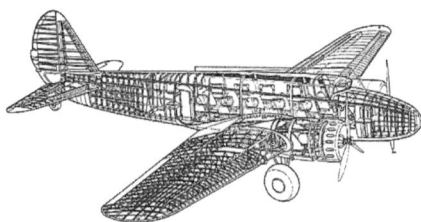

a)   b)

Fig. 2.45 a) *Boeing 247 (1933)* [F25]; b) *The fuselage structure* [30]

Keller and Bechereau have built an airplane where the spar was placed in the leading edge of the elliptical Aluminum wing. All ribs were attached to this spar [28]. In 1934, Duncanson used a circular aluminum tube as a wing spar and in 1936 R. Vogt used the steel tubular spar, containing also fuel [28].

Fig. 2.46 *Hawker Hurricane – the central section* [F26]

In 1935, Hawker Hurricane had a non-conventional structure: the Warren beam fuselage with longitudinal steel longerons and bracings by Duralumin, attached with clamps (Fig. 2.46). Over a secondary wooden structure of ribs and stringers was placed the fabric skin. The wing was a Warren beam, covered with fabric, in 1939 a metallic stressed skin being added (Fig. 2.47); the airplane is an example of transition to monocoque structure [23].

The geodetic construction has been used at Vickers Wellesley bombers (1935) and Wellington (1936) (Fig. 2.48 a). The structure consists in a rectangular system of diagonals which are coiled around the fuselage and four longerons aligned with the fuselage axis (Fig. 2.48 b). In references [28] and [47] are described the advantages of this type of structure: "For each load spectrum, a diagonal will be subjected to tensile and the other to compression with the same values, at any intersection point of the diagonals system. The tension tends to reduce the curvature radius of the diagonal while the compression tends to increase it. The two effects are balanced at the intersection points, having as result the stabilization of the compression and of the bending moment of the diagonals."

Although, the material was effectively used resulting a well-proportioned structure, the constructive complexity has led to large manufacturing costs [28], [29]. In [47] the geodetic construction is characterized as "extremely rigid, and can withstand to the damages caused by anti-aircraft projectiles." Due to the technological difficulties (Fig. 2.49), this type of structure was not further developed; after the 1960s, the concept was re-evaluated, but only for the space industry.

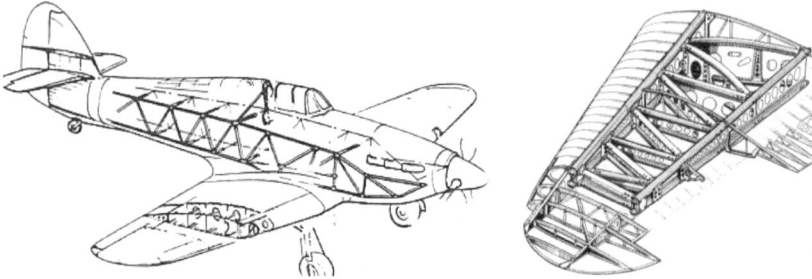

Fig. 2.47 *Hawker Hurricane (1935) with the Warren beam structure* [23]

In 1936, it was distinguished the Douglas DC 3 passenger airplane (Fig. 2.51), the one of the first successful metallic airplanes [12], symbol of the 1930's [60]. It is considered that the DC 3 is the first airplane that became commercially profitable [35]; between 1936 and 1945, there were built 11.000 units (the airplane is still in service; according reference [47], in 2002 there were still in service around 1000 units!).

A special construction of the airplane's wing presented Junkers Ju 87, this having a big number of transverse ribs and stringers (Fig. 2.50). J. Northrop built the multi-cell tri-spar wing and independently by Rohrbach, he discovered the lack of the structure failure due to the local buckling [02], [30].

The Northrop wing was adopted at the first Douglas DC aircrafts, in the case of DC 3, the model having good performances and a long operational life [02]. DC 3 used the concept developed for the Northrop Alpha, having in addition the upper side of the wing stiffened with corrugated sheet metal [47] (Fig. 2.52). The DC 3 success demonstrated the superiority of the concept with stressed skin, frames and stringers made of riveted Aluminum; after the 1930s, this arrangement became the preferred concept [47].

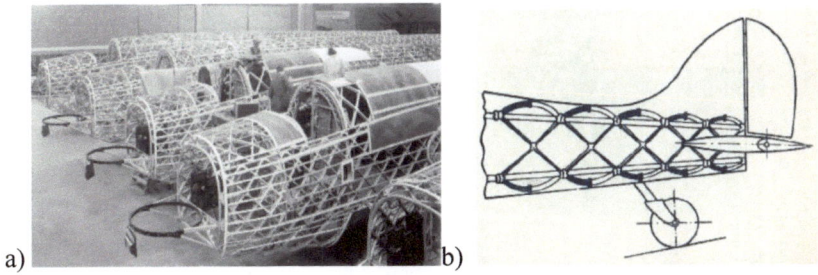

a)   b)

Fig. 2.48 a) *The Vickers Wellington bomber (1936)* [F13];
b) *The layout of the geodetic spirals* [24]

Fig. 2.49 *The Vickers Wellington bomber fuselage* [F27]

Fig. 2.50 *The Junkers Ju 87 airplane wing* [F28]

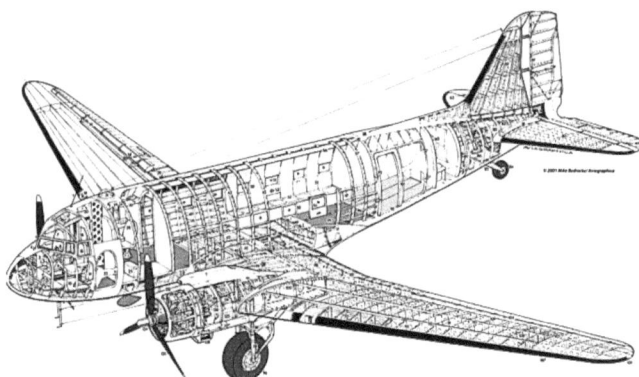

Fig. 2.51 *Douglas DC 3 (1936)* [F29]

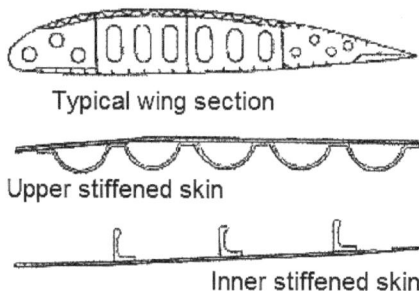

Typical wing section

Upper stiffened skin

Inner stiffened skin

Fig. 2.52 *Detail of DC 3 wing section* [47]

Quoting Howe, [47] he attributes the special success of DC 3's ("capability and performances beyond the dreams of designers") by:

- a solid basis, founded on the experience of previous aircrafts;
- a rational and scientific approach of the new introduced concepts;
- a careful testing program of the first aircrafts, with the integration of feedback in the subsequent generations.

It should be noted that in the design teams it is stated that: "the airplanes are made from other airplanes", which shows that an airplane that would compete with the existing ones on the market is an impossible objective without previous experience with other aircrafts. Often, an aircraft is derived by the fuselage lengthening or by cross-sections increase, searching for the greatest possible recovery of what already demonstrated a good behavior. It can be asserted that the airplanes are made with "save as" even if this novelty percentage varies, or in some situations there is no matter of a "template".

The B247D passenger aircrafts and DC-3 employed an Aluminum structure; over this structure it was riveted a sheet metal skin. Except for the material, this concept was identical with the one of the Albatros aircraft, which had the plywood bonded or nailed over the wooden frame [30]. Although, Ford Trimotor was fully metallic (1920) only Boeing 247D (1933) and Douglas DC-3 (1935) were the aircrafts which have definitively demonstrated the superiority of metal construction, presenting many similarities with current passenger aircrafts [30].

The main stages of the metallic wing development are (fig. 2.53):

- the Junkers wing with Pratt beam longerons with tubular flanges, diagonal bracings instead of ribs and corrugated self-stiffened skin;
- the Dornier wing with Warren beam longerons with tubular flanges, ribs from sheet metal with holes for weight reduction and stressed skin; the leading edge ribs are used;
- the Rohrbach wing with longerons and ribs with sheet metal webs with holes for weight reduction and stressed skin; the trailing edge presented beam type ribs. The leading edge is stiffened by bracings;
- the Northrop multi-cell wing (the rectangular cells were between the longerons core and ribs); the holes for weight reduction have formed edge. The ribs present vertical stiffeners; the skin presents stringers that prevent the local buckling.

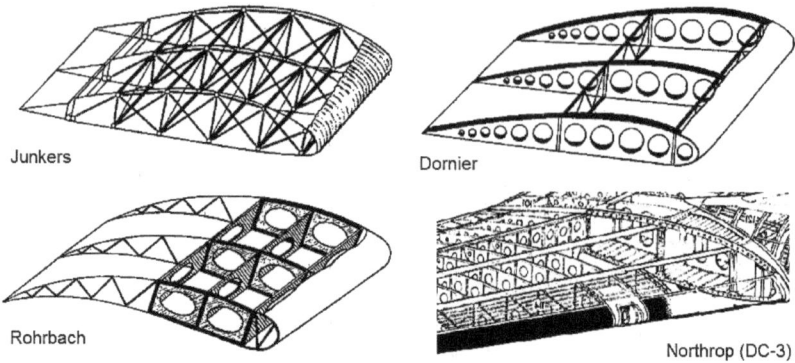

Fig. 2.53 *The stages of wing box development* [F21]

In 1937, Lockheed X 35 was successfully tested, the first aircraft with pressurized cabin [60], allowing the flight at high altitudes. The effects induced by pressurization under the structure would be known later.

In 1939, Beech D17 S was build, the one of the last biplanes with wooden wing, welded frame, wooden shaped frames and fabric skin.

The full metallic structures made of Aluminum with stressed skin, of monocoque type have become a standard in the period of maturity of the propeller aircrafts. The integral "vegetable" airplane has become a representative of a bygone era [02].

The transition from the latticed beam to monocoque has been a slow process, which was overlapped with the transition from wood to metal. The final was the transition to metal monocoque.

2.6. THE 1939 – 1945 PERIOD

Employing a twin fuselage, Lockheed P38 Lightning (1939) was the first airplane with flush head rivets [35], in order to increase the quality of the aerodynamic surfaces. Along with this advantage, the flush head rivets generated a set of problems, such as:

- the knife effect (for thinner metal sheets) which produces the shear of cone shaped head;
- the drilling of the chamfered holes, influencing the rivet head forming.

Fig. 2.54 *The twin fuselage airplane - P38 Lightning* [F30]

The higher speed was one of the requirements that still dominated the military aircrafts market. The speed may be increased by increasing the engine power or by lowering the drag. Because more power requires heavier engines and more onboard fuel, the reduction of drag was one of the priorities. The researches in this area led to laminar airfoils with superior aerodynamic characteristics to the existing ones [35]. The maximum thickness point of the airfoil has been moved from 25% of the chord to 30 ÷ 40%, allowing the placement of the longeron close to the neutral axis, thus improving the wing stiffness [35].

At the beginning of the World War II, the lack of qualified workers, machinery and duralumin led to orders for wooden airplanes (except for the fighter aircrafts) [28]. The new wooden aircraft (including the Mosquito bomber) had new adhesives (synthetic resins) with superior water and fungus proof, and also adhesives allowing the wood - metal bonding [28]. The pre-compressed wood impregnated with resins started to be used, in some applications being fastened with screws [28]. Another reason for using wood was to involve in the war effort the big number of available qualified carpenters.

Efforts to replace metal, led to a lot of combination in materials. Tests were performed on surfaces made from resins without additional wood, but the poor results have led to the need for a "fibrous" material [28]; this problem would be solved later with fiber glass.

Despite of the many technological improvements, the wood was further used only for gliders [28]. For the infiltration into enemy territory, a special category of gliders was designed, with a big payload, such as CG3 in USA or Hamilcar in England, with a useful load of 16 tons (Fig. 2.55).

Fig. 2.55 *The wooden glider - Hamilcar* [F31]

In this period, the wooden sandwich structure started to be used, consisting of two plywood sheets and balsa wood core [28]. Firstly, this was used in the Sunset flying boat floaters (1919) and then for the Karman/ Stock glider fuselage (1924). In [47], it is mentioned that the designers' intention was to perform a real frame reinforced coque structure.

The sandwich structures have entered into a cone of shadow, being used only in 1938 for the construction of the four engines passenger airplane De Havilland Albatross [47]. The Salunier Morane 406 airplane (1938) had the skin made from a material called Plymax, a sandwich structure with wooden core plated with an Aluminum sheet [11]. The idea has only become attractive, later at the end of war, when it was found out that de Havilland Mosquito (released in 1941, Figure 2.56) had the fuselage made from sandwich structure (spruce veneer with balsa wood core, with a total thickness of 11 mm) [28]. In the areas where the structural elements were attached (concentrated forces) the balsa wood core was replaced by spruce blocks [47]. However, Mosquito had problems with humidity and microorganisms from the tropical environment, all these affecting the quality of these glued joints [47]. For this reason, many planes are disintegrating in flight.

Fig. 2.56 *The fuselage of the De Havilland Mosquito airplane* [F32]

Since the 1940s, different experiments with sandwich structures with Aluminum, steel or glass fiber skin with the core of cellulose acetate, polystyrene, polyvinyl or rubber have been performed [28]. The Mosquito's success has stimulated the research for finding a mixture of materials suitable for the sandwich structures [28]. For the Supermarine Spitfire combat aircraft (1943) some phenolic matrix composites, reinforced with hemp fiber were used in the wing longeron and in the fuselage components. The wing longeron flanges were made from concentric "U" profiles; their length gradually decreasing to the wing tip (Fig. 2.57).

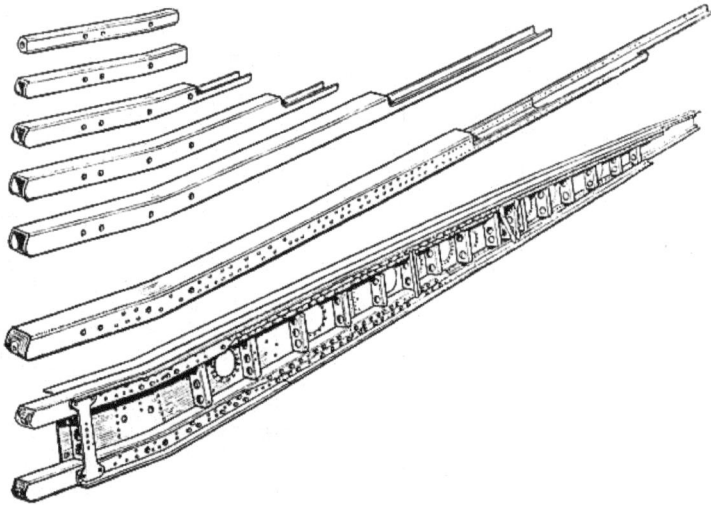

Fig. 2.57 *The wing longeron of the Spitfire airplane* [F33]

As a curiosity, it is worth mentioning that in 1947, the Hughes H-4 Hercules airplane (with a wing span double in comparison with a Boeing 747) was made entirely from wood (the main reason being to preserve the Aluminum, from strategic considerations) [35]. Having a maximum take-off weight of 180 tons (!) after the first flight, the airplane was grounded. In Figure 2.58 the monocoque fuselage with reinforced web frames and bracings for keeping the perpendicularity of the frames to the skin may be observed. In Figure 2.59 it observes the latticed beam structure of the trailing edge.

Fig. 2.58 *The fuselage of the Hughes H-4 aircraft* [F34]

Fig. 2.59 *The trailing edge of the Hughes H-4 aircraft* [F35]

The war years is distinguished only rom the large production series, the aircrafts design being contradictory. The wood was combined with metal; the monocoque structure was combined with latticed beam. One of the priorities was for the quick repairing (Junkers J 87 have a modular structure to allow for fast repairing by replacement).

The metal was preferred because it allowed higher speeds. However, the plywood was still used for a long time, together with the fabric (especially for control surfaces). The metal was used in a wide variety of semi-products and assembly methods.

Specific for this period, is the emergence of many configurations determined by the diversity of the missions (long-range bombers, faster interception-fighter aircraft, aircrafts with short takeoff, high capacity gliders). The increased speed and height required stiffer structures, with stressed skin and monocoque structure as prevailing solution. The thinner aerodynamic airfoils required new solutions for the wing stiffness.

In a very short time the large series production led to the need for more productive assembly methods. In [47], it is mentioned the research of the Curtiss-Wright for an innovative stitching method for the sheet metal parts, which, combined with riveting and spot welding, was able to reduce the manufacturing time. The solution could not be implemented, later being reconsidered by NASA after the 90s, for composite structures.

2.7. THE 1945 – 1985 PERIOD

This period is dominated by technical solutions requested by the supersonic speeds and heights up to the stratosphere limit. From the structural point of view, the engineers had to understand the fatigue behavior of the metallic structures, obtaining airframes with up to 30 years life time. The progress has been made regarding the sandwich structures and the implementation of new assembly methods.

De Havilland Commet was the first passenger airplane with pressurized cabin (1952). After the loss of three aircrafts and after the investigations problems caused by fatigue (especially due to the stress concentrators introduced by the square passengers' windows) have been found. After the structure upgrade, it came back in service in 1958, but it could not cope with the competition (Boeing 707 and Douglas DC-8) [12], [35].

Fig. 2.60 *The Boeing 377 Stratocruiser fuselage section* [F36]

Fig. 2.61 *The Breguet Atlantic aircraft with double lobe fuselage*

In the 1950s, more producers (Breguet, Curtiss) migrated to the double lobe fuselage (bubble section). This configuration had the role to increase both the passengers and cargo compartment (Fig. 2.60, 2.61). The fuselage skin is stressed only to tensile load for the circular section. Additionally, at the bubble section, the bending moments occurs, which leads to a heavier fuselage for the same pressurization loads. This configuration is more advantageous in the case of unpressurized cargo compartment (such as the bomb compartment at the bombers). The configuration has not been used after the 1960s, the last airplane of this type being the Hawker Siddeley Nimrod.

In 1966, the American B47 bomber (1947) was withdrawn due to fatigue problems of the primary structure. The problems started to be solved only with B52 (1952) [35]

In this period, the freighter aircrafts were being developed, having a similar structure as the passenger' aircrafts, but in addition it had the reinforced deck, upper cargo door and no windows.

According to [23] and [35], in this period, efforts were made to:

- reduce the number of rivets by returning to the bonded structure or by the manufacturing of structures which integrate multiple components into a single one;
- understand the fatigue effects and cracks occurrence and propagation by building structures that can be easy inspected or redundant structures with multiple load paths;
- reconsider the role of light structures along the aerodynamics and the engines within the overall efficiency of the aircraft.

Based on the Comet's experience, Boeing 707 (1958) incorporated into its structure concepts as fail safe and damage tolerance [12]. B707 set the layout for passenger airplanes that lasts until our days [60].

If the fighters of the World War II had the airfoil relative thickness of 14-18% [35], in order to conquer the supersonic speed newer thin profiles (4-9%) were needed [35]. Starting in 1960s, NASA tested different stiffening systems for the missiles boosters. The special requirements for tightness have led to the use of the integrally machined skins. After the successfully use of the orthogonal stiffening aligned at 45 degrees to the fuselage axis, in 1964, NASA launched a program to identify the geometry of a structure with optimum behavior for compression loads [74]. The studies indicated that the best behavior of structures with triangulated cells [74]. Based on the best behavior are those of the triangulated beams, their layout was extrapolated to the cylindrical structures [74]. The new structure named "isogrid" (Fig. 2.62) is distinguished by the [74]:

- very good torsion behavior;

- many points where external loads can be attached in the structure nodes (such as thermal protection or solar panels);
- an overall strength to allow the maneuvering on the orbit.

Although the idea of a system of stiffeners different from the one where they are aligned with the fuselage axis is older (Vickers Wellington), the new technology developed by McDonnell Douglas and NASA has not became attractive for the aircraft producers due to the high manufacturing costs, this remaining only in the area of spacecraft.

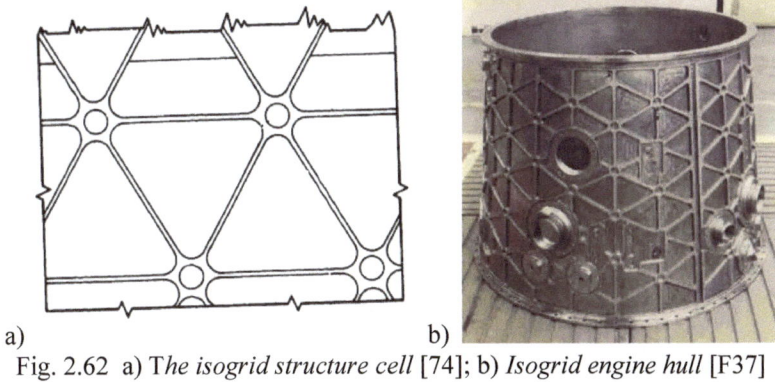

a)   b)

Fig. 2.62 a) T*he isogrid structure cell* [74]; b) *Isogrid engine hull* [F37]

The development of the supersonic aircrafts led to a shorter wing, of relative small thickness [28], multi-longeron, with thick skin [12]. One of the first successful airplanes that used this configuration was the Dassault Mirage III (Fig. 2.63).

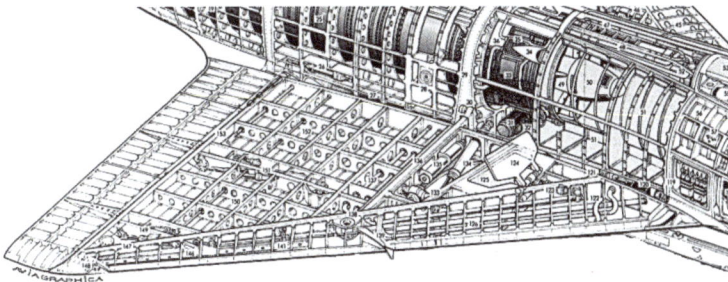

Fig. 2.63 *Dassault Mirage III - The delta multi-longeron wing* [F38]

For the aircrafts exceeding the speed of 2 Mach, skins made by sandwich materials with steel honeycomb (good behavior at temperatures above 150° C) were used; one example is the Convair B58 airplane (Fig. 2.64), made in 1956 [02]. Also called "the bonded bomber", it had Aluminum panels that covered 90% of the wing and the 80% of fuselage [47]. Due to the sandwich structure, the structure weight was only 16.5% of the maximum take-off weight [47]. During the operations, the aircraft presented problems with the strength of the bonded elements [35]. At Convair fiber glass was extensively used [35]. In [47] it is mentioned that the special maintenance problems led to very high costs (for the wing these were up to five times higher than for the B52 bomber wing, which had two times larger surface). Also in [47] it is stated that although there is no data about the structure behavior in time (due to the withdrawal after only 10 years in service), it can be supposed that the lifetime was also limited due to the problems raised by the sandwich structure.

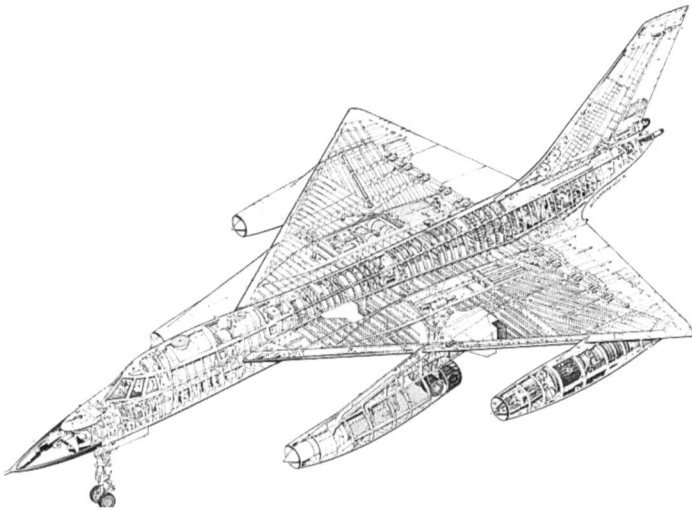

Fig. 2.64 *Convair B58 –Skin by sandwich materials* [F29]

The sandwich structure (steel brazed honeycomb) has been extensively used on the B70 Valkyrie (1964) - 64% of the entire structure weight [47]; in 1968, the extensive operated C5 Galaxy used the bonded honeycomb structure for most of the secondary structure: leading edges, slats, trailing edges, flaps, nacelles, floor panels, engine pylons, fairings and the main landing gear box [47].

The honeycomb structure was made from two Duralumin sheets (0.5-1.0 mm thickness) with a honeycomb core made from thin sheet-metal strips (approx. 0.2 – 0.3 mm thick); all were bonded by an adhesive film. Having a very good bending strength to weight ratio relative to the other structures, many producers considered this as the structure of the future.

Fig. 2.65 *Examples of insertions for the honeycomb panels* [F39]

There are also drawbacks for the honeycomb panels; one of them is the requested attachments on the whole contour with a maximum pitch of 100-120 mm. Due to the fragility of the honeycomb core, mounting on the panel cannot be made in threaded holes or by bolting. The honeycomb panels' producers have found an adequate solution, designing dedicated insertions (Fig. 2.65). To install an insertion, first a cutout in the upper sheet is made, followed by the honeycomb core trepanation. After the insertion is placed using a coordination hole drilled in the lower sheet, the cutout in the upper sheet is plugged with a bonded sheet metal patch. In the end, all parts are bonded by injection of polyurethane foam. The insertions of this type allow the attachment using screws or bolts. There are also special insertions that allow the attachment of seats (seat tracks), equipment or the attachment straps for the cargo containers.

Machined frame with corrugated core

Extruded frame

Reinforced machined latticed beam frame

Removable latticed beam frame

Machined frame with vertical braces

Integral machined latticed beam frame

Latticed beam frame machined on the both sides

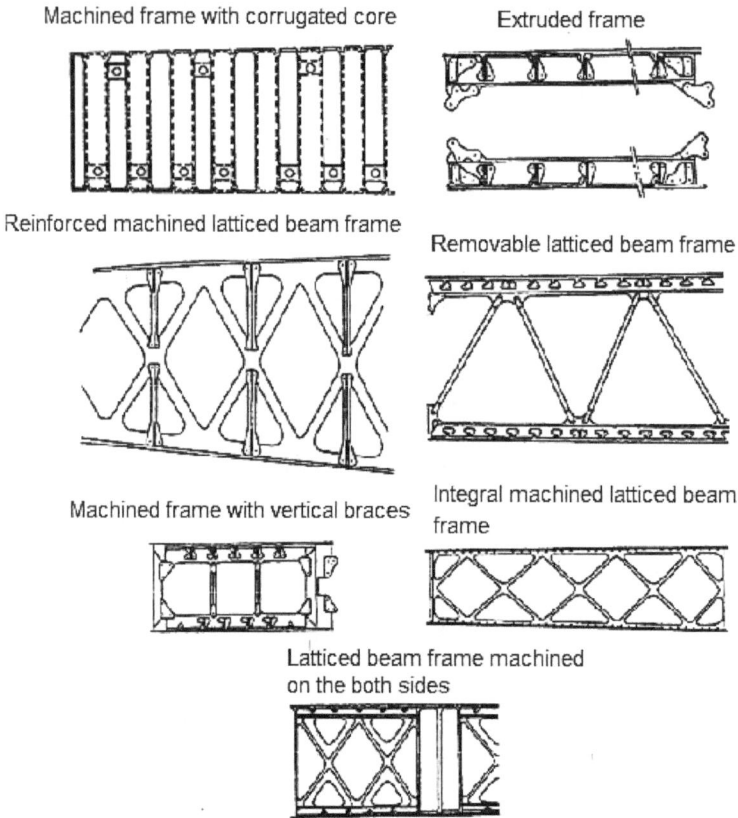

Fig. 2.66 *The Concorde airplane - details of latticed beam frames* [F40]

The insertions allow the attachment of loads of hundreds of pounds while the efforts are dispersed on the whole panel surface, maintaining the cohesion between its components.

One of the most remarkable airplanes of this period was Concorde. Because a supersonic aircraft haven't been built at that size, the designers had to solve a lot of problems raised by the supersonic speed. A first measure was to design the whole fuselage body from a single material (Duralumin 2618A) in order to prevent the problems of stresses and strains due to the thermal deformations.

For the thermal stresses lowering the decision to use the latticed beam was made. In Figures 2.66 and 2.67 the structural details of the ribs and the longerons of beam type, made in a multitude of constructive shapes can observed. In the area of fuel tanks, the membrane walls have been manufactured by machining on both sides, obtaining a corrugated section [11].

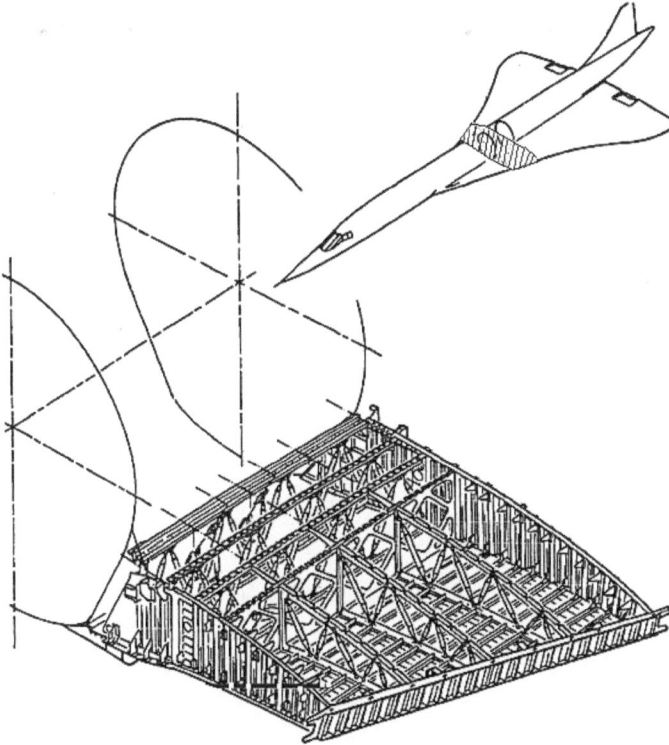

Fig. 2.67 *The Concorde airplane - The wing section* [F40]

Although the obtained weight was satisfactory, the repairing of such structure has proved to be very difficult. The fuselage had a classical construction, while the wing was ingeniously split. It was segmented as the fuselage barrels, being made common sections of wing and fuselage (Fig. 2.68), jointed with steel bolts (Fig. 2.69). Although the designers had to solve a multitude of new problems, the fuselage behavior under the mechanical and thermal fatigue conditions were as expected; the airplane had been for 27 years in service with a fleet of 20 units

Fig. 2.68 *The Concorde airplane – joined fuselage-wing section* [F40]

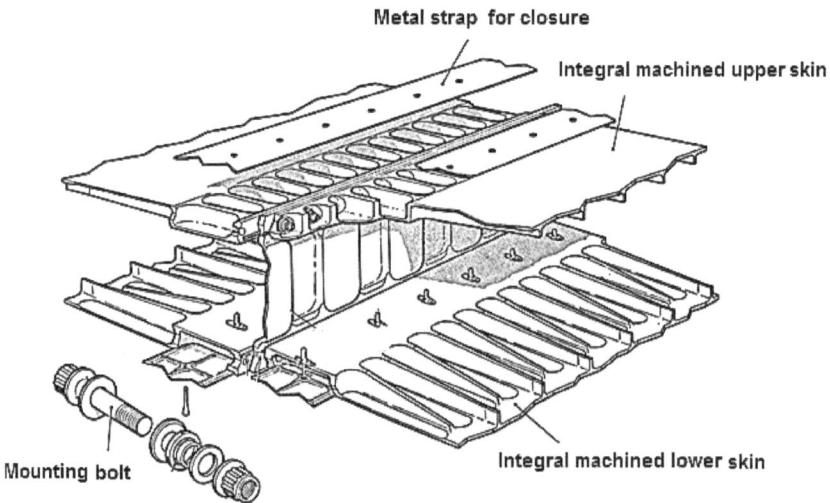

Fig. 2.69 The *Concorde airplane – Detail of wing sections jointing* [F40]

Another new feature was the realization of large machined parts. As a result of the trend to reduce the workmanship and the number of components of the assemblies, the opportunity of the big machined parts has been studied. Beside the development of the milling machinery and technologies, the manufacturing of an integrated machined panel resulted more economic than its riveted equivalent. Initially, only the flat parts were machined (from plate raw material) for the wing ribs, bulkheads, junction frames, etc. An example for that time (1963) was the rear bulkhead of the lunar module designed by Grumman Company (Fig. 2.70).

Fig. 2.70 *The rear bulkhead of the lunar module LM-2* [F41]

The large machined parts started to be used for the wing longerons, engine pylons, fuselage frames, or fuselage engine frames. Another innovation was the skin milling for weight reduction and the distribution of the material thickness according to the intensity of the stress flows. The chemical machining (chemical milling) consists in the controlled removal of a uniform material layer of a metallic part by immersing it in a chemical vat.

The separation of the areas to be removed by erosion from the other areas is done by the protection of a compound. Used since the 19th century, as a method of artistic etching, the process was patented in 1956 by the North American Company as a method for structural weight reduction for the missiles. This procedure has rapidly spread; to most aircrafts with thick skin (over 2.5 mm) using this process for the fuselage or wing skin (Fig. 2.71).

The chemical milling was drawn very well as a replacement for the mechanical milling, in the case of large parts with complex shapes (lengths of 5-10 m with double-curvature surfaces). These could be processed only by using large machinery, with a minimum of 4 axes. For large sized parts, especially those with thin walls, the mounting, aligning and milling raise some problems regarding the stability and vibration, requiring low speed cutting to prevent the deformations

The chemical milling allowed the manufacturing of very accurate pockets for weight reduction, with a low level of energy consumption. Some companies have been caution in implementing these technologies due to the material fragility as a result of the chemical corrosion. In order to obtain a hardening of the chemical milled surfaces, shot peening may be used. Shot peening (with lead balls) was also used to form the curvature of the wing skins as an alternative to cold forming.

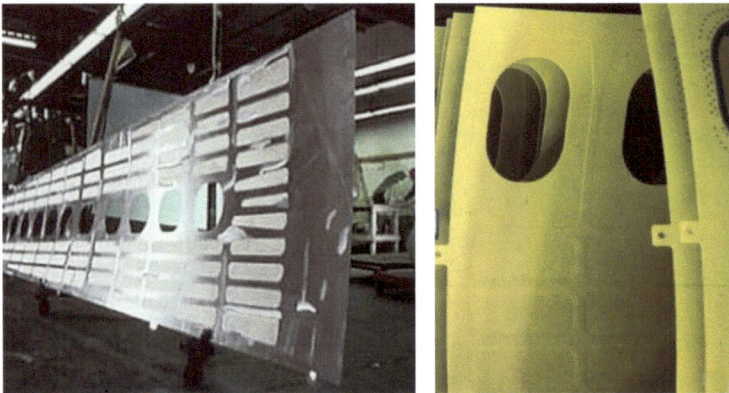

Fig. 2.71 Chemical *milled fuselage skins* [F42]

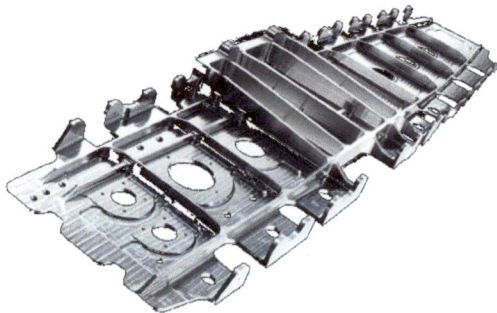

Fig. 2.72 *Wing rib (1600 × 375 × 70 mm)* [F43]

The milling had stabilized for most modern aircrafts as a solution for large and stiff parts (Fig. 2.72). The large milled parts of Duralumin are met in the joining areas of the wing, stabilizer, fin, landing gear, etc.

In order to reduce the structural weight, one method was to use hybrid structures, consisting of assemblies with parts produced by different technologies or materials. An example is given by the space shuttle (1980), where for the large frames a unique combination of milled frames, riveted structures and latticed beam were used. The beam bracings were made by aluminum-boron alloy (Fig. 2.73).

Fig. 2.73 *The space shuttle hybrid structure* [F44]

Another method for weight reduction was the design of deep milled parts, followed by their assembly into a torsion box. The Tornado airplane had the central torsion box made of titanium, having the pivots of the variable swept angle wings at the extremities. As an integrally milled part, it embedded the walls of the torsion box, the stiffening ribs and the fittings lugs (Fig. 2.74).

After 1960, the weight reduction researches brought in the forefront the composite materials. The boron fibers in epoxy resin matrix were initially studied. Subsequently, composites with glass fiber, carbon or Kevlar (for ballistic protection) were developed.

The F4 Phantom military aircraft (1960) had the fin made from boron fibers; in 1974, F14 Tomcat had the stabilizer made from metal honeycomb core (throughout the airfoil thickness with the boron fiber skin). The F16 airplane (1978) had already about 50% of skin form carbon fiber, with a layer of glass fiber for the inners [Niu, 1992]. A particular feature of the empennage structure was the stabilizer with the staggered longerons (Fig. 2.75).

Fig. 2.74 The *Tornado airplane – the torsion box by milled titanium* [F18]

The F18 Hornet airplane had a 12% structural weight realized thanks to the carbon fibers [44]. For the AV-8B Harrier airplane (1985), the Boeing Company made for the first time the wing internal structure from composites. The multi-spar wing had a corrugated web to increase the buckling stability (Fig. 2.76).

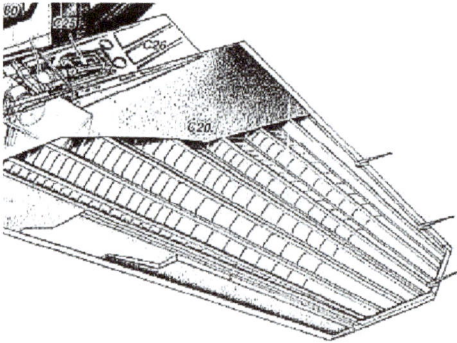

Fig. 2.75 *The stabilizer with staggered longerons (F16 Falcon)* [F44]

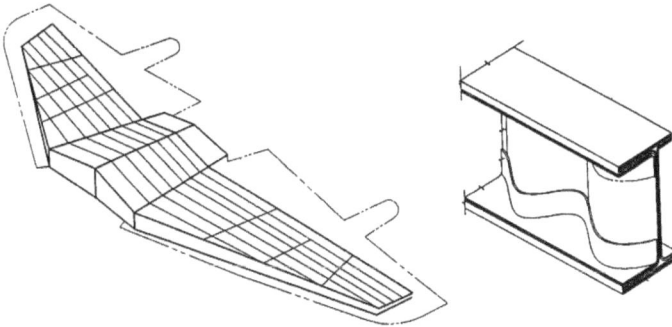

Fig. 2.76 *Multi-spar wing with corrugated web (F16 Falcon)* [43]

The first composite structures had very high manufacturing costs. For this reason, the decision to make a composite structure in place of a riveted aluminum structure was justified by approx. 30% weight saving, 40 ÷ 60% reduced number of components and 50 ÷ 70% less fasteners [43] (Fig. 2.77).

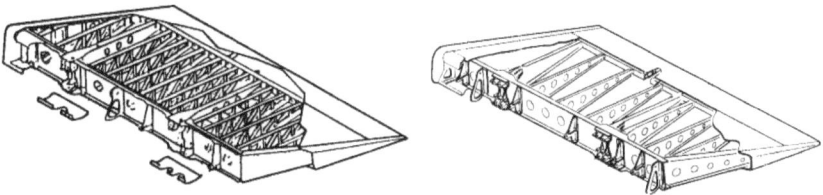

Fig. 2.77 *Aluminum vs. composites aileron (The L-1011 airplane)* [43]

An important leap in the introduction of composites in the aircraft structures was made by NASA within the Aircraft Energy Efficiency program (ACEE), targeting the design, manufacturing and testing of the composite structures. Firstly, there were considered only the secondary structures (the aileron of the Lockheed L-1011, the stabilizer of the Boeing 727 and the rudder of the McDonnel-Douglas DC-10). In the second phase, composite primary structure components were considered (the drift and the stabilizer caisson). The main criteria applied in the design of composite structures were [43]:

- the parts must withstand to the same loads as metal structures;

- the efforts transmitted to the adjacent structures shall be identical or lower than those of the equivalent metal structures;
- the mounting interface on the aircraft must be the same for both metal and composite structures;
- the aircraft maneuverability must not be affected (especially the response to flight controls or the occurrence of aero-elastic phenomena).

Fig. 2.78 Composite *fin, typical for Airbus aircrafts* [F45]

The experience gained with the ACEE program led to a growing trust and an increased percentage of the composites for the structural elements. The composite structures with honeycomb core also started to be used on a large scale. Boeing extended composites for the new generation of aircraft (B747, B757, B767, B777) from the empennage and ailerons to the spoilers and flaps. Airbus was the first producer which implemented the composite structures for the series aircraft (the Airbus A310 fin). The approach was gradual, starting in 1972 with the fin leading edge from made glass fiber for A300, in the 1980s, extending the technology to aerodynamic brakes and landing gear doors [43]. For A310, Airbus finally got a 22% lighter fin, with only 95 parts vs. 2076 parts of the Aluminum structure (Fig. 2.78).

Lear Fan 2100 was the first airplane that had an integral composite structure (1981); presenting a weight saving of 40% in comparison with the aluminum structure. The fuselage had the skin of 0.35 mm thick, withstanding the loads induced by pressurization too [43]. The fuselage was made from bonded overlapped panels. The tri-spar wing had the skin with integrated spar flanges. The spar web had a honeycomb core to withstand the shear loads (Fig. 2.79). Despite of many inovative features, only three aircrfts were built without having the type certificate from the FAA, the reason being the lack of reliability of the central gear.

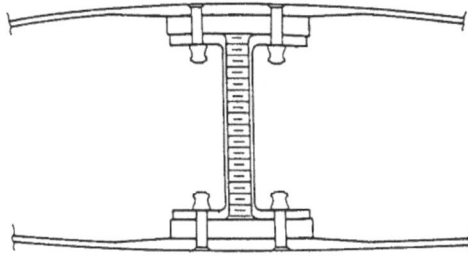

Fig. 2.79 *Lear Fan 2100 – The wing spar with honeycomb core* [43]

The Star Ship airplane (1983) had the skin made with sandwiched carbon fibers and honeycomb core, presenting a limited number of frames and stringers. The fuselage consists from two monocoque halves with variable thickness from 12.7 to 25.4 mm, depending on the stress value. The wing was made from a sandwich structure, the spars and skin panels being assembled by means of aluminum angles. The airplane was produced in a limited number of units due to the following disadvantages of the sandwich structures [23]:
- the forming of single-curvature surfaces leads to unwanted areas with double curvature;
- the water infiltration and condensation lead to corrosion and the unbounding of the core.

The composite structures have been implemented on the fuselage only partially, in the non-structural areas (access doors, hatches or fairings). One of the reasons of the reluctance to use the composite materials for the fuselage was the lack of information regarding the behavior on fatigue. Although accelerated fatigue testing can be performed in the laboratory, the true feedback comes only from the field of operation, where data is collected from structures that have been loaded under real conditions. The behavior during long periods of time is not affected only by mechanical loads but the combination between these and corrosion, vibration, humidity, temperature gaps, etc.

Fig. 2.80 *The radome of a F111 after a bird impact* [F46]

The composite structures have not been agreed because of poor impact behavior. Because of the absence of the plastic range, after a local impact, a composite material skin will not absorb the impact energy, presenting an extensive failure; thus, a local repair will not be possible. An example is given by the impact with birds, a risk that although is managed by all airports, it is not yet eradicated. In figure 2.80 the total failure of the radome (nose fairing) of an F111 airplane due to a bird strike is shown.

The Speed Canard airplane has introduced a number of innovations such as the glass fiber skin with PVC foam core [43]. The spar's flanges were made from unidirectional carbon fiber (UD). For the first time, the main landing gear struts were made from carbon fiber (outer side) and kevlar (inner side). The wings were mounted in a central section by the insertion of the spars into a rectangular beam; the fixing was made by a centering stud and a mounting bolt.

The period after the World War II was marked by the research in the field of the structural layout, materials and manufacturing technologies. The milling process reached maturity and the composites made progress marked by design and materials diversity. The riveted structures were optimized and an important progress was made to the pressurized structures. Titanium starts to be used for the high-speed aircraft structures, subjected to excessive heating but also for fasteners (screws, rivets, lock-bolts, hilocks, etc.). All these progress allowed for new speed, height or endurance records, that still remains unbroken. During this period, extreme airplanes like the Lockheed SR 71, Concorde or Voyager were developed, that will remain forever as milestones in the aviation history. Remarkable progress has been made in terms of expanding the operational period, civil aircrafts (Boeing 737, 747, Airbus A310) or military (F14, F16, F18, Puma, Lynx, Alouette) proving to be extremely long, being still in activity after 40-50 years from the date of the prototype flight.

2.8. THE 1985 – 2000 PERIOD

The Italians from the Piaggio Company continued to implement the composite in the airplane structure, using the sandwich structure with glass fiber for the empennages, control surfaces, engine nacelles, nose cone of the fuselage, wing leading and trailing edges. Thus, composites reached 20% percent of the structural weight. As a feature adopted by Piaggio, the fuselage has the longitudinal section with an aerodynamic airfoil shape (Fig. 2.81); this idea was put into practice since 1924 by Vincent Brunelli with the RB-2 flying machine. Although the fuselage lift leads to a smaller stressed wing surface – as well as to a lighter wing - the concept is impractical for the internal volume and from manufacturing considerations (the cylindrical fuselage has most of the frames identical).

There was a tendency to replace the honeycomb with foam, but these panels were heavier and with poor mechanical properties. The honeycomb structure was used for the floor panels, helicopters blades and fairings (with no structural role).

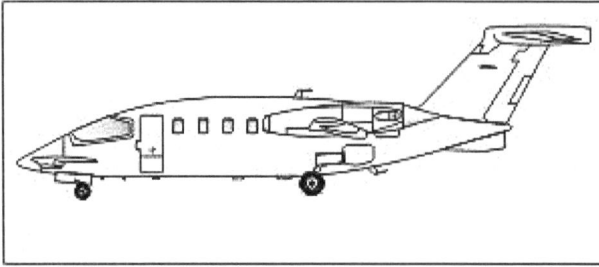

Fig. 2.81 *The Piaggio P180 airplane with lifting fuselage* [F47]

In 1987, Burt Rutan demonstrated the superiority of the honeycomb composite structure (Hexcel) with the Voyager airplane, having a structural weight of only 10% from the maximum take-off weight. In [47] it is mentioned that one of the reasons for the use of composite honeycomb was to decrease the number of parts and the manufacturing costs compared to the traditional riveted structures. The skin was made from 0.36 mm thick carbon fiber, with honeycomb core of 6.35 mm. The wing structure (Fig. 2.82) was made up two monocoque halves, incorporating the fuel tank too. The closed shape of the wing torsion box generated a three-spar wing, with a main longeron and two secondary longerons. The leading edge and the trailing edge have no structural role, being attached to the wing box, by means of the external ply by carbon fiber. The central spar has the flanges made with unidirectional carbon fiber. After the wing halves bonding, the access to the fuel tank was no longer possible. The wing had the carbon plies oriented at ± 32° relative to the flight direction, which gave a good torsional stiffness but were to elastic for bending loads (Fig. 2.83) [43].

The fuselage had a skin structure similar to the wing. To reduce the structure weight, the firewalls of the two engines also had the role of structural stiffeners for the fuselage junction with the wing and front wing (the canard stabilizer) [43]. For the same reason, the main landing gear was located at the joining between the wing and the fin beams, the latter having the role of the fuel tank too.

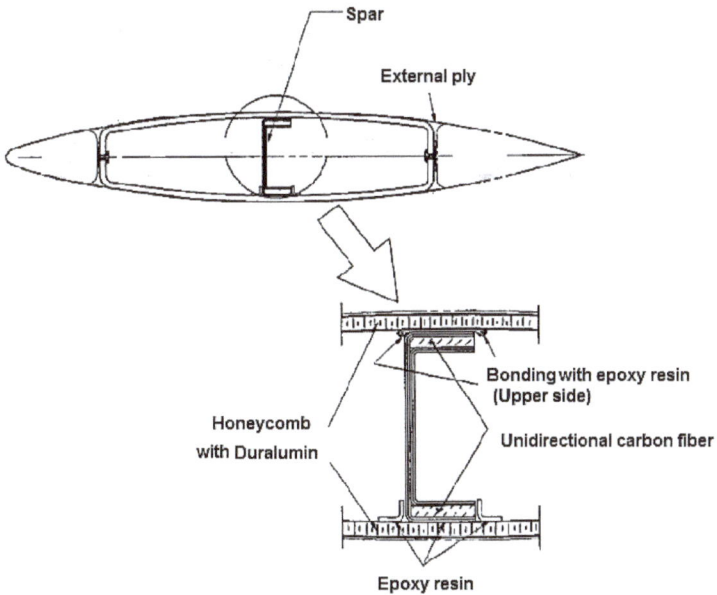

Fig. 2.82 The *Voyager airplane wing* [43]

Fig. 2.83 *The Voyager airplane* [F48]

In 1989, Boeing performed its first flight with the V22 Osprey that had 50% of the weight structure made from composites. The composite wing had spars, ribs and skin with integrated stringers. The empennage had spars with corrugated core. Also, in 1989, the B2 bomber was the largest aircraft with full composite structure (170 tons).

The fighter aircraft fuselage was mostly developed based on the high speed aerodynamic considerations (small aspect ratio wing), but also to facilitate the access to equipment (large openings in the fuselage). In [47], it asserts that their structure can hardly be called monocoque, because of the fuselage openings; thus the bulkheads and the frames take over a great part of the aerodynamic loads.

To understand the difference between the producers and the evolution of the structural layout of the airplanes (layout) after the Second World War, dedicated researches have been made. These targeted the pitch of the structural stiffening elements of the fuselage and the wing (frames, ribs), and the number and location of the spars in relation with the chord of the wing airfoil, respectively. The conclusions presented in [58] are as follows:

- the structural layout of the commercial aircrafts is very similar regarding both the time periods and the producers;
- for military aircrafts large variations from one producer to another are observed (for example, for the short wing, some prefer a large number of spars and a small number of ribs, while others do the opposite);
- for the commercial aircrafts the trend was to keep the pitch of the stiffening elements constant, while for the military ones, the tendency is to vary the pitch of the structural elements depending on the size of the local loads.

It can be asserted that the metallic aircraft had reached a stabilization of the structural arrangement, which indicates a certain level of maturity.

2.9. THE 2000 – PRESENT PERIOD

In the early 2000's, in reference [47] is stated that aircrafts have been a hybrid between metal and composites, approximately being manufactured in equal percent from aluminum, titanium and composite materials. Regarding the metallic parts, the progress achieved by the manufacturing technologies (the 4 or 5 axis machines, with large ranges, high accuracy and spread working regimes) has accelerated the implementation of the integrated construction. Thus, the large-sized stiffened panels with longitudinal and transverse elements could be built from monolithic parts, with a manufacturing time and cost smaller than the equivalent riveted structures. Another advantage of milled parts was the repeatability, identical parts being produced with a minimum human intervention. Also, the quality control is much faster, because once the numerical command program is certified the risk of non-compliance is much lower.

Fig. 2.84 *The fuselage junction of the Airbus A400M fin* [F42]

Until the 1990's, milled parts were avoided, being used only where they could not have been replaced with riveted or welded assemblies. They were considered expensive, difficult to manufacture and with a reduced use of the raw material (at least 80% of the raw material was removed). The designers who were against these milled parts, called them "sculptures". A lot of today's aircraft structural parts (pressurized walls, heavy loaded frames, junction fittings etc.) are made extensively by milling of Duralumin (2024, 6061, 7075 and 7050) or titanium alloys; the percentage of the remaining material can be down to 3% from initial volume (Fig. 2.84).

With the improvement of topological optimization software, the massive flat parts (especially the wing ribs) have uncommon shapes, which look more like latticed beams. In figure 2.85 two examples of milled leading edge ribs (nose ribs) from of the Airbus A380 airplane are shown.

Fig. 2.85 Airbus *A380 – The ribs of wing leading edge* [F49]

After the year 2000, the expansion of composite into the primary structure of the large airplanes is observed. Airbus A400M (2009) has composite wing skin, with aluminum ribs and metal fuselage (Fig. 2.86). Airbus A380 (2005) has a metal wing (with composite flaps) and a hybrid fuselage (Fig. 2.87). For the first time, this was made from carbon fiber (rear side) and Glare (composite with glass fiber and duralumin) for the front and rear fuselage. The glass fibers were used only in small percentage (wing drop panels and fin leading edges).

The most recent large civil aircrafts - Boeing 787 (2009) and Airbus A350 (2013) replaced the Aluminum skins, for the first time building most of the fuselage and wing from carbon fiber. Thus, the last phase of the transition from metal to composite, or from Duralumin to carbon fiber was accomplished. By comparison to other components of the primary structure, the fuselage has additionally the pressurization load and the passengers' proximity.

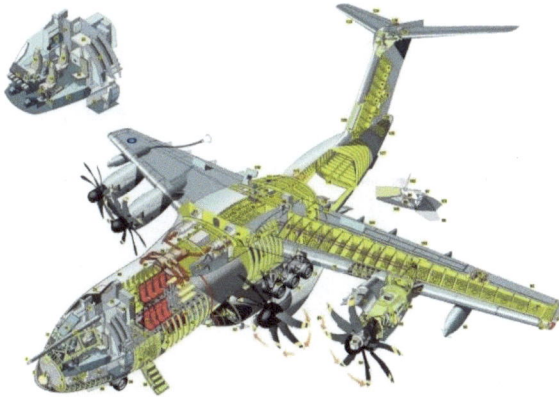

Fig. 2.86 *The Airbus A400M aircraft structure* [F50]

Fig. 2.87 *The Airbus A380 materials shares* [F10]

For this reason the big producers have decided that the fuselage would be the last major component to be made from composite.

The manufacturing technologies of the fuselage are different, Boeing uses winding (with epoxy resin impregnated wires around a turning cylinder), while Airbus has remained at the fuselage panels, assembled by riveting (Fig. 2.88). If the transition to a new material stimulated the development of new technologies for parts manufacturing, in aviation, the riveting still remain the main assembling method. Although the bonding or curing or the integrated structures (in autoclave) are known for since the 60's, the solid rivets with their derivatives (Hi Lite, Lock, Lock Bolt, etc.) are used in all heavy loaded or fatigue joints.

Fig. 2.88 Airbus *A350 – fuselage panel and the rear bulkhead* [F45]

In modern applications, the glass fibers are used less. In the beginnings, these were used because of the affordable manufacturing cost. After the carbon fiber production became cheaper, glass fiber were the first to be completely replaced. The glass fiber is still used in areas where a higher elasticity is required or where the small loads don't justify the use of the carbon fiber. The glass fiber can be used as a pad between the carbon fiber structure and the aluminum parts (to prevent corrosion), or in drilling areas, to prevent the destruction of the external layers of the carbon fiber.   The kevlar applications are restricted to areas where the impact with external objects is possible or for the armor panels.

The carbon fiber is progressively used more in areas made until now exclusively from metal. Thus, parts belonging to mechanical assemblies (levers, shafts, transmission) or even gear's housings are experimentally made from carbon fiber. The trend is to replace many categories of metal parts with carbon fiber. Many of these parts are only at the demonstrators' level; only in time the most effective materials and technologies for a particular category of parts will be settled.

Although, after the 1980s the isogrid structures have entered into a cone of shadow, the research started again at NASA and Hexcel, then at Alkan, without any implementation in production. In 2011, Isogrid Composites Canada reported the development of a technology that succeeded in realizing isogrid carbon fiber panels, with better characteristics then Hexcel and Alkan. Isogrid Composites announces weight reductions up to 50% to the aluminum panels, for an increased strength of 30% (Fig. 2.89). In [47] it is stated that these structures have very good damage tolerance to low-speed impacts.

Fig. 2.89 Isogrid *carbon fiber structures* [14]

According to Composites Forecast & Consulting USA, the composite aerospace structure production has recorded a worldwide spectacular increase for the period 2010 - 2015, with 100% growth (Fig. 2.90). These data indicate that the technological progress of the composite materials has passed the critical threshold, the focus now being on the increase of the industrial production.

NASA is currently studying an older assembly technology – the composite sewing - based on the Advanced Composite Technologies program started in collaboration with Boeing in 1990. In [31] it is presented the PRSEUS - Pultruded Rod Stitched Efficient Unitized Structure concept (unitary efficient structure with extruded stitched bars) with the objective to develop a wing concept after 2016. In figure 2.91 it is presented an exploded view of the main components of a monocoque structural joint, consisting of:

- frames made from foam (1), wrapped in carbon fiber (2), with cutouts for stringers (2a);
- doubler strips for frame - skin joint (3);
- carbon fiber stringers (4) with roving rods (5);
- doubler strips for stringers - skin joint (6);
- carbon fiber skin (7);
- stitching lines (8) of the components.

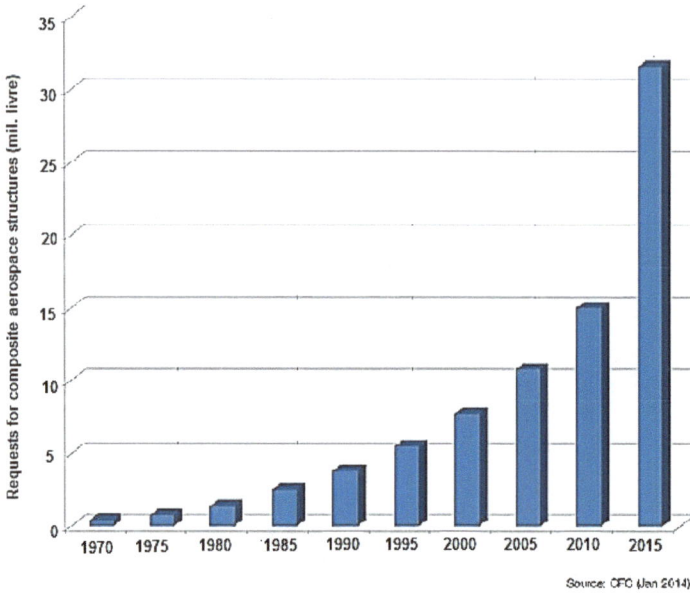

Fig. 2.90 *The request of composite structure* [F06]

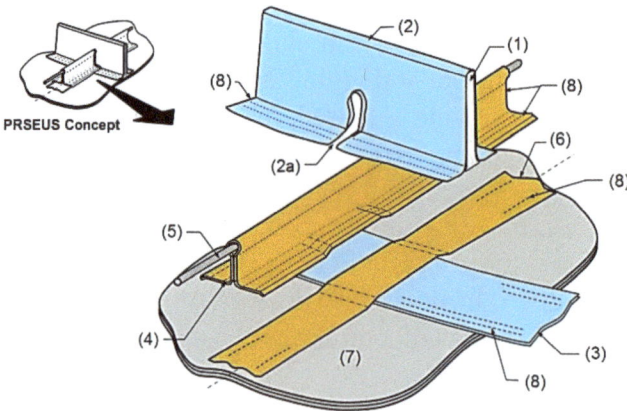

Fig. 2.91 *NASA/ Boeing PRSEUS concept* [31]

As for aerodynamic configurations, there is no significant news. Researches have been made to achieve a unique body for wing and fuselage (Massachusetts Institute of Technology), respectively the airplane with variable wing "adaptive wing" (NASA), but these are only in the stage of concept (Fig. 2.92) [23].

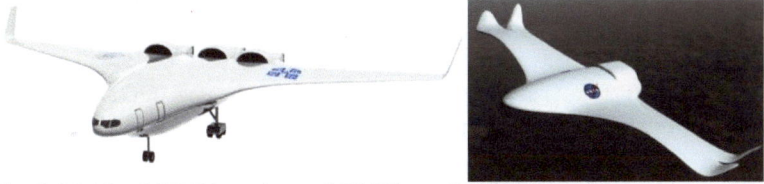

Fig. 2.92 *The MIT Silent Aircraft* [F50], *and NASA Morphing Aircraft* [78]

As research directions, the following may be identified [47]:

- Multifunctional structures - structures that, in addition to the strength and stiffness requirements, include some features related to the survivability, thermal and aerodynamic efficiency combined with constraints as low-cost manufacturing, maintenance and repair;
- Structures that contain sensors for Structural Health Monitoring and actuators for the modification of the mechanical parameters for an optimum behavior at different vibration modes;
- Structures that correspond to the integrated concept of reduced operating costs, by increased time period between inspections, increased damages monitoring capacity and on request maintenance and repairing;
- The integration into the structure of the antennas, the electronic countermeasures and the radar-absorbing systems;
- The virtual prototype (Simulation Based Prototype) – allowing an evaluation since the project stage, of the functionality, maintenance and aerodynamic requirements.

As advantages, the virtual prototype allows:

- Testing of the pilot and operators interaction with the aircraft (by simulation of the flight and the maintenance operations using the virtual reality) for the project validation and the shortening of the training time;
- The analysis of the different versions and project configurations;

- The simulation of the part manufacturing, assembling operations, including the ergonomy of the workplace;
- An easy interface with the manufacturing systems;
- The real - time programming of maintenance/ repairing operations of the in service aircrafts, even in the situation of structural damages occurrence.

Another priority is to develop affordable composite structures, identifying the following areas of research:
- The sewing and braiding of the composite structures to prevent the detachment and delamination problems;
- The insertion of reinforcements to prevent the occurrence of cracks and to improve the impact strength;
- Composites combining different materials (carbon, kevlar and glass fibers);
- Innovative methods of jointing between composite and metal (by reeling, using channels, by tapering, etc.)
-

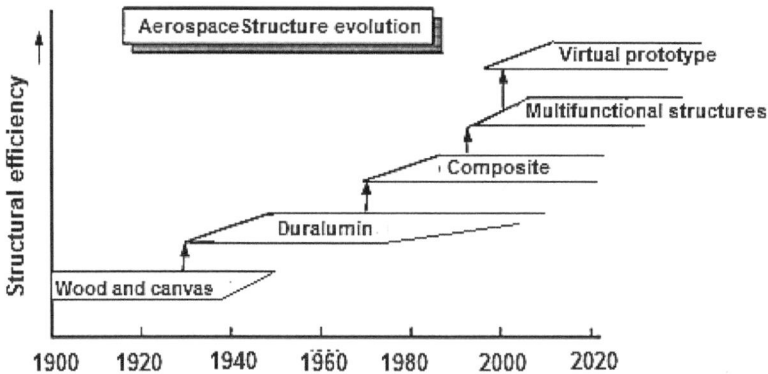

Fig. 2.93 *The main evolution stages of the structural efficiency* [30]

According to reference [47], the multifunctional structures and the development of the simulation methods like the virtual prototype will realize a significant leap forward in the structures efficiency (Fig. 2.93). It can be observed that, until year 2000, the increase in efficiency was achieved mainly by the introduction of better materials, with appropriate manufacturing technologies of the parts and assemblies. After year 2020, the composite structures will gain a new dimension, surpassing the simple role to withstand to mechanical loads, becoming an integrated system. In addition to this new feature, the revolution brought in simulation by the improved hardware and software tools will allow a higher accuracy in structures design and sizing. Thus, a visible improvement of the aircraft efficiency is expected for the next decade.

2.10.   Conclusions

The structure design played an important role in the aircraft's evolution. "*The repeated compromise between strength and minimum weight, made the airplane's structure one of the most optimized*" [12].

In [30], it is stated that the aviation history is rather evolutionary than revolutionary. There are authors who consider certain aircrafts as revolutionary. However, looking back, it can't be stated that a revolutionary aircraft has triggered a chain reaction which generates a wave of significant changes. Thus, there are airplanes as Deperdussin, Zeppelin Staarken, Junkers J 1 or Doriner Wal, which, despite the fact that they may be considered as revolutionary; they had a modest impact on the producers. An explanation might be the great diversity that characterized the aviation world in the first 20 years, and the lack of theoretical knowledge; a remarkable flying machine could hardly stand out in a very dynamic environment, without having some solid comparison criteria. From another point of view, it was hard to consider an aircraft as revolutionary because the aviation progress was made with very small steps, even though apparently things have moved very fast.

In a very dynamic world, it took more than 20 years for the monoplane, the monocoque or duralumin to get full rights. On the other hand, the information circulated very fast: the problem of standardizations was raised since 1918 - most of aircrafts were biplanes, with bi-longeron wing, with similar wing structure, the latticed beam fuselage, etc.

In the attempt to point out a few reasons for the weak impact of the reference aircrafts, the following considerations are identified:

- from the first years, the producers were oriented to the immediate or short term advantages; the long-term advantages were not understood due to the limited theoretical knowledge and the lack of know-how about the operating costs;
- a revolutionary aircraft doesn't mean the default commercial or operational success; the success is the one that draws attention not only to engineers, but also to the investors and owners who are able to initiate the development of a flying machine;
- the development of an aircraft was a risky business and any investor preferred the small steps;
- the early-stage development of materials and the manufacturing technologies has not offered support to the industrialization of the revolutionary concepts.

The highly successful aircraft had either a limited innovation degree or just have incorporated many concepts introduced by other producers. From other perspective, some extraordinary airplanes as Zeppelin Staarken were destroyed in 1922 from political reasons, and the aircraft without propeller built by Henri Coandă was considered at Le Bourget in 1910 only as "La curiosité du salon".

CHAPTER 3

Structural concepts

It took almost 30 years since the first controlled flight for the aircrafts to reach to the structural and aerodynamic configuration which, today, is considered "classical". The aviation pioneers searched for different structures for wing, empennages and fuselage, respectively, which involved considerable research and development efforts. The monocoque configuration with monoplane wing is common to the most airplanes. If the latticed beam is still used, the biplane was abandoned after WWII.

In this chapter are presented, by manufacture and functional point of view, the main structural concepts for wing and fuselage, completed with comparisons, trends and comments.

3.1. BIPLANE AND MONOPLANE

The monoplanes were the first flying machines used, in order to perform various missions; because of a multitude of longerons and tensile wires to increase the wing strength and the stiffness, these structures became complex and the aerodynamic performance remained poor (Fig. 3.1). Since the first years, the biplane was a great success compared to the monoplane – the take-off distance was *shorter* and the wing mass was lower [50]. The monoplane was considered unreliable versus the more robust biplane [35]; being considered at that time as "of low strength and dangerous" [12]. For instance, in 1913, the UK Ministry of Defense exclusively ordered biplanes [64]. Furthermore, the authorities from many countries had hardly accepted or even prohibited the use of monoplane [12]. A clear disadvantage of the monoplane was considered the reduced visibility that is critical for the survival in the aerial battles.

Fig. 3.1 *The monoplane – looked with distrust in 1910 [F52]*

From the structural point of view, the biplane (the overlaid wings with the system of struts and tensile wires) constitutes a high beam with high moment of inertia, with a superior behavior at bending and torsion compared to the monoplane [23], [12], [49]. The lift of an airplane is proportional to the wing surface (the projection of the wing on a plane). Generally speaking, a biplane allows a wing surface on a span of only 50 %, relative to the one of a monoplane, therefore, more robust. Actually, this is 71% [50]. Therefore, it was agreed the idea that a biplane facilitates a lightweight and stiffened structure [23].

In order to increase the lift for a more compact wing, several flying machines with overlaid wings have been realized in that period, known as multi-planes. It was also built a concept having no less than 20 lamellar wings (Fig. 3.2), which remained only in the demonstrator stage. The only successful multi-planes were the tri-planes, which in the 1918 – 1919, dominated the fighting aviation [25]. These were preferred against the biplanes, due to the increased lift of a compact airplane, despite of the increased drag. Being developed during the war, the capacity of more onboard ammunition and the maneuverability were prevailing. In addition to these, the wing had the lower span, leading to smaller bending loads, thus the structure being lighter.

Fig. 3.2 *Multi-plane airplane (1904) [F53]*

Afterwards, it was demonstrated that the monoplanes had better aerodynamic characteristics. It was proved that the induced aerodynamic drag (which appears at the wing tips, due to the air circulation from the lower to the upper side) represents an important component of the drag. On the other hand, the wing with the high elongation is the one which offers the best lift to drag ratio. Not the least, the overlaid location of the two wings produces the decrease of the efficiency, due to the aerodynamic interferences [02]. Therefore, the monoplane with the big span is superior to the short wing biplane, with four areas inducing aerodynamic drag and a low wing efficiency. Before 1918, it was considered that the lower wing may only have 90% of the lift of the upper one, due to the "unfavorable influence" of the biplane configuration [50].

An important aspect is revealed by [02] – the wings of the first airplanes were so thin, that these couldn't support the aerodynamic forces, requiring two wings stiffened with pillars and tensile wires. It is considered that the biplane is still preferred for the acrobatics due to the strong wing, which allows higher loads factors than the monoplane. The first fighters wouldn't have cope to the inertial loads occurring in an aerial battle (close steering with fast speed) with a monoplane wing. Also, due to the reduced span, the polar momentum of a biplane is lower than the one of the monoplane, allowing faster spin, thus more aggressive maneuvers.

Nowadays, the biplane is reconsidered by a research teams from the Tohoku University from Japan and the Massachusetts Institute of Technology (Fig. 3.3). Based on the concept of the German engineer, Adolf Busemann (1950), the delta wing biplane revealed significantly reduced sonic bang on the virtual prototype simulations, which could re-launch the passengers transport in the big towns' area.

Fig. 3.3 *Supersonic Biplane with joint wings [F54]*

The modern aircrafts use exclusively the monoplane, the biplane being dropped out after Second World War. The monoplane wing had in its turn, a separate evolution, starting with the trapezoidal wing, then the broken wing (seagull wing), elliptical wing, swept wing, delta wing or the wing with variable arrow. Actually, the swept wing equipped with high lift devices and winglets (small wing tip fins), is the widest accepted solution.

3.2. LATTICED BEAM AND MONOCOQUE

The latticed beams structure was the first structure that fulfilled the minimum stiffness and lightweight requirements on the Wright brother's flying machine - Flyer 1903. The earlier solutions were considered unsuccessful due to the fragility (i.e. the gliders of Otto Lilienthal or the aircraft of Samuel Langley) or too heavy.

The primary structure represents that part of an aircraft structure, which, leads to a catastrophic event upon failure. As a central element of the aircraft structure, the fuselage has a significant number of primary structure components; for that reason, its design, calculation and testing are regarded with a special care both by the producer and the certification authorities.

Fig. 3.4 *Wooden latticed beam fuselage [F55]*

The first structures used on the early airplanes were the latticed beam for fuselage, and the stiffened skin for the wing (containing radial stiffens like the bat wing). Related to the fuselage, this met the following stages:

- latticed beam without skin, consisting of 3 or 4 longerons, with vertical bracing and diagonal tensile wires (Fig. 3.5);
- beam with columns, stiffened with tensile wires and external fabric (Fig. 3.6 a);
- beam with columns with tubular bracing and external fabric;

141

- beam with partial stressed skin (veneer or plywood); it had a reduced number of tensile wires and bracings (Fig. 3.5);
- beam with intermediate structure as support for the skin (Fig. 3.6 b);
- beam with stressed skin;
- mixed fuselage (latticed beam - on the front, monocoque - on the rear);
- coque fuselage (Fig. 3.7 a);
- monocoque fuselage (Fig. 3.7 b);

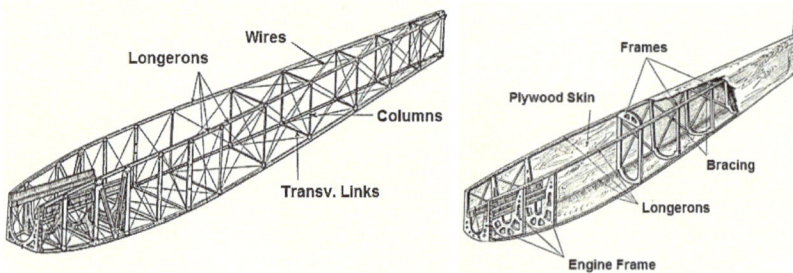

Fig. 3.5 *Latticed beam and wooden latticed beam with skin fuselages* [50]

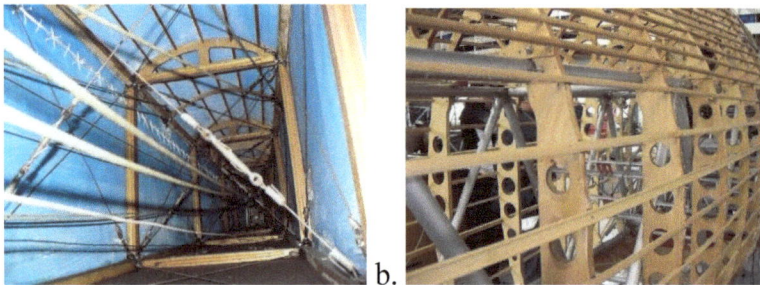

a.   b.

Fig. 3.6 *a) Sopwith fuselage [F56]; b) Beam with secondary structure [F57]*

The coque structure consists of a single stressed skin; this is not used in aviation due to the poor behavior to buckling that led to the large wall thickness and an unpractical weight [28]. The frames reinforced coque refers to the stressed skin reinforced with frames and bulkheads; this was firstly used on wooden airplanes. The monocoque represents the stressed skin stiffened with a supplementary network of stringers in order to maintain the thin skins stability (the most common on modern airplanes). According to [23], the monocoque term is defined as "the frames reinforced coque, with discrete discontinuities".

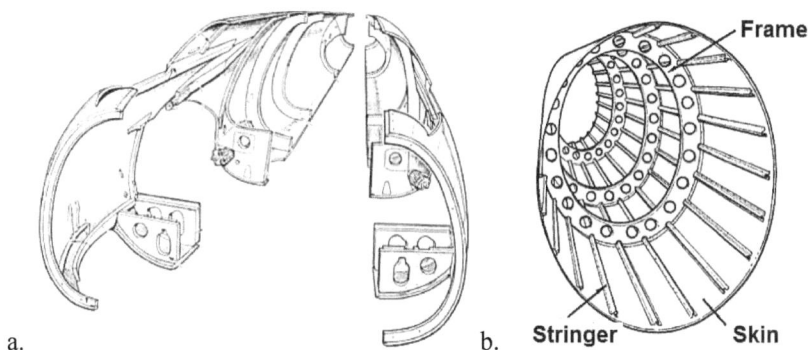

Fig. 3.7 a) *Frames reinforced coque [F58]; b) monocoque fuselage [F59]*

In the English literature, only the "monocoque" term is used. In Cambridge Aerospace Dictionary, the definitions of frames reinforced coque and monocoque are identical:

Frames reinforced coque = tridimensional shape (as the fuselage) having all the strength at the level of skin, frames and adjacent stringers, without an internal structure or bracings.

Monocoque = structure where the loads are partially supported by the combination of frames and stringers, partially by the skin.

The thin walled structures where the skins support the loads are named stressed skins. According to other sources, the frames reinforced coque structure is considered to be the one where the skin is the main stressed element, being supported by the structure with frames and stringers, the monocoque being the structure with force stringers (longerons) with the main role in supporting the loads.

According to [23] and [02], it is accepted that the technique of monocoque structure concept was originated from the ships industry (yachts construction), these two domains intersecting each other in the area of the flying boats manufacture. In 1911, the Swiss engineer, Ruchonnet was the one who designed the first monocoque structure [47]. However, in 1912, the monocoque was simultaneously used for the first flying boat (Curtiss USA) and for the race airplane (Deperdussin). Being an immature technology at that period, the monocoque was reconsidered with the metallic skin of the flying boats (1924, Rohrbach). In 1921, F. Loudy, patented a monocoque metallic structure with swept skin, as shown in Figure 3.8 [47].

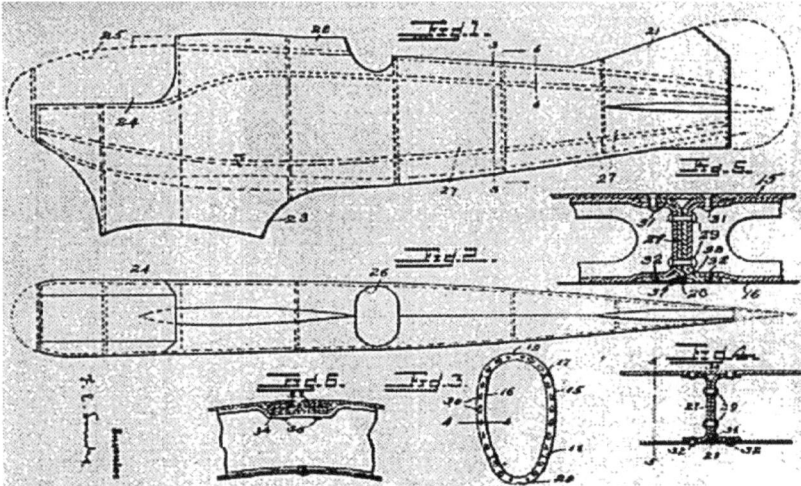

Fig. 3.8 *The patent of F. Loudy for the monocoque structure (1921) [47]*

The required power for the propulsion of an aircraft depends of the cubic speed. In order to keep the power at a minimum level, at the high speeds, it is important to keep the cross section at a minimum level and to maintain an aerodynamic shape as good as possible [28]. The rectangular section of the latticed beam requires an outer swept surface, to minimize the aerodynamic drag (Fig. 3.9. b). The monocoque offers in a single structure the solution to answer both requirements: the aerodynamic shape and withstanding to all loads [28], [23]. On the other hand, the monocoque presents a more uniform stress distribution [28], reducing the stress concentrators and leading to a better fatigue behavior. The material is placed at the surface of the section; both inertia and the polar momentum are bigger, the bending and torsion strength being better for monocoque [23].

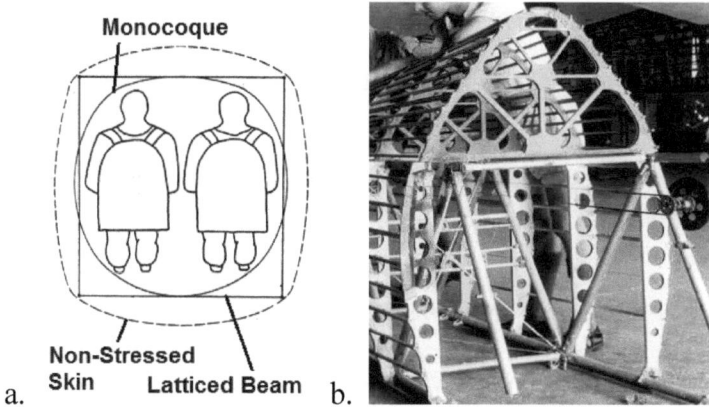

Fig. 3.9 *a) Latticed beam vs. monocoque sections [F28];
b) beam with secondary structure carrying the skin (Hawker Hurricane) [F60]*

As against of the latticed beam with skin, the monocoque allowed of an approximately 35% smaller cross section, leading to lower drag and less power (Fig. 3.9 a) [28]. As consequence, the monocoque uses better the inner space [49]. According to [50], it is specified that the monocoque was "unpractical", even if it was a great success by aerodynamic point of view [32]. The monocoque presents the problem of stability to compression and shear stresses, these loads leading to the most common failure modes (Fig. 3.10). In order to increase the critical buckling forces, stringers were added and the frames pitch was decreased [23].

The latticed beam offers certain advantages such as the ease inside access and configuration, due to the flexibility of its constructive elements arrangement [29]. A particular advantage has the welded latticed beam, which provides an integrated solution of beam with the joint fittings for the connection with the main structural components (wing, empennages, landing gear, etc.).

Fig. 3.10 *The loss of the stability of fuselage skin [F60]*

Fig. 3.11 *The structure reinforcement for the passengers door [F61]*

It is mentioned that on thin – wall structure, each hard point (mounting point) requires an area of gradual decrease of the skin thickness; the joint can be made only in the area presenting stiffened structural elements, such as the frames. The monocoque is vulnerable at the large apertures in the fuselage skin, as those for the passengers or cargo doors. For this reason, these must be special reinforced, so that the stresses will be redirected to the adjacent areas, and the stiffness of the new structure will maintain the fuselage pressurization (Fig. 3.11).

The latticed beams with members assembled by riveting require a large number of diagonal elements, the internal volume being practically unavailable. Figure 3.12 shows the Spitfire fuselage, where the riveting was adopted in order to use Aluminum instead of Steel, for obtaining a lightweight structure.

Fig. 3.12 *Riveted latticed beam fuselage [F62]*

The big drawback of the latticed beam comes from the stresses concentration, the failure of an element leading to the considerable decrease of its strength [29].

According to [23], the adoption of the monocoque structure naturally coincided with the transition from wood to metal, the transition being made once it was possible the obtaining of cheaper sheet metal raw materials of a good quality of the surface and high mechanical characteristics. According to [30], the transition to metal was only done after the monocoque structure advantages were already proved, in two stages, firstly by replacing of wood with metal and then, in a subsequent stage, by adopting certain constructive solutions for the pieces and assemblies, appropriate to the manufacturing   technologies specific to metal.

The transition process to metal started for the Germans constructors, in 1916 (Junkers); the essential contributions were brought by Dornier (1918) and Rohrbach (1919), targeting first the durability and the removal of the problems generated by the wooden constructions. The transition ended, with some exceptions, before the Second World War. The wood was kept for limited application in the case of the gliders, until 1970, when it was completely replaced by metal or glass fiber.

The transition to monocoque started in 1912, this solution finally gaining against the latticed beam, after 1935. However, the beam structure was used in the helicopters industry (till '50 – '60), continuing to be used on a large scale on light and ultralight aircrafts, where the monocoque development, manufacturing and production preparation costs were leading to no-effective products, by commercial point of view.

The advanced idea presented in [30] is sustained by the fact that through the transition from metal to composite, the structure doesn't record constructive novelties, the monocoque having the same components, shapes or predominant assembly methods (the riveting and bonding). The big commercial airplanes producers adopted the composite pressurized fuselage, very late (after 2010). Nevertheless, the structural elements are identical to the metal structures ones: stressed skin, frames and stringers assembled by riveting.

Reconsidering an older idea of the isogrid structure, within the FP7 project - Advanced Lattice Structures for Composite Airframes, the researchers from England, Germany and Russia investigated new possibilities to combine the advantages of monocoque structure with isogrid structure, using some methods of topological optimization and genetic algorithms.

The new structure was named non-isogrid or latticed fuselage, the stringers being coiled, with a higher density on the ceiling and in the floor area, the areas corresponding to the windows being free (Fig. 3.13). The researchers consider that the results of the study may be implemented with minimal changes to reach to an acceptable preliminary project.

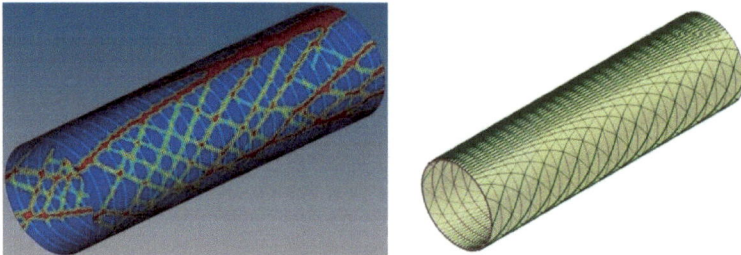

Fig. 3.13 *Latticed fuselage [53]*

According to its long-term strategy, Airbus intends to develop a bionic fuselage, with a cellular structure to allow a panoramic view. The idea of this structure is inspired by the bones, which simultaneously offers good strength and lightweight. Such a structure would become a spatial latticed beam, subjected to pressurization with transparent panels replacing the windows.

3.3. STRUCTURAL CONCEPTS - CONCLUSIONS

Although, similar to materials, during the time, the aircrafts structure knew many configurations, only the ones fulfilling the operational and costs requirements finally won. From all above, the following conclusions may be drawn:

- the monoplane was employed as an aerodynamic and structural solution, only when the materials and the assembly technologies allowed the realization of an enough robust structure;
- the monoplane isn't competed on short - medium term. Some authors consider that the actual aircraft has the configuration of a bird - mono-fuselage, monoplane, with the empennage located on the rear. As long as the design is close of the nature's models, the seeking of different concepts may seem irrational;
- the latticed beam and the monocoque are still the most usual concepts for fuselage;
- the latticed beam with rectangular cross section will disappear due to the fact that technology to produce carbon fiber fuselage is more and more affordable;
- the actual research directions are oriented to continuous mediums (foams, cellular structures) or discrete environments (latticed fuselage or cellular fuselage);
- the most advanced actual projects (Boeing B787 and Airbus A350 XWB) adopted the classic monocoque structures, this is a new reason to consider that the commercial aviation is

extremely pragmatic and it goes on only with small and safe steps;

- a significant leap forward is expected from the light airplanes area or the automotive industry where a lot of resources are invested for the development of light innovative structures and the appropriate affordable manufacturing technologies.

CHAPTER 4

The evolution of aerostructures design and calculation

If the flying machines had a very interesting evolution, the way in which these were designed and calculated is also worth to be investigated. In the beginning of aviation, the individual producers built the aircraft directly, without any previous design or just by improving previous aircrafts. The only aircrafts for which a minimal technical documentation exists are those which were patented. If the aircraft was involved in military operations, it also had the operation, maintenance and repairing handbooks.

Technical descriptions at engineering level were made in dedicated magazines, *Flight Global* and its supplement *The Aircraft Engineer & Airships* being a reference in the domain. Alongside the images, manuals and books, magazines represent valuable sources for the reconstruction of many forgotten pages of the technical history of aviation.

The first project plans were the templates for the airfoil and sections, requested by the sketching of the contour line on the sheets of plywood and by the need to manufacture identical parts. With the development of the aerodynamic shapes, the definition of surfaces became a very qualified work, where using only planar curves, complex surfaces were defined (Fig. 4.1).

The simple bent surfaces (for lifting surfaces) were the easiest to build by having as reference the airfoil and its scaled contours. The surface of the fuselage was generated starting from some central sections (usually the engine area and the cockpit) and the intersection curves from the symmetry and the median horizontal plane. By intersecting the surface with parallel planes with the transversal section, the frames outlines were obtained; sketching these on the drawing board was an operation of an incredible complexity.

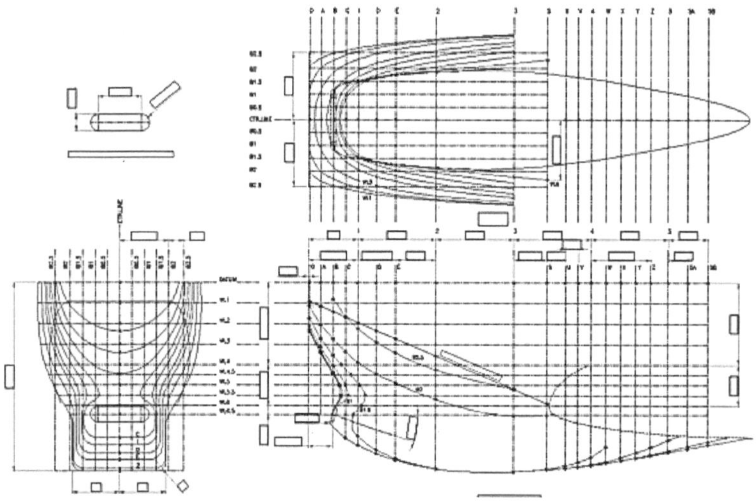

Fig. 4.1 *The definition curves of the surface of Spitfire's radiator inlet [F63]*

The evolution of the sections along the longitudinal axis was not continuous for the first airplanes. The engine compartment defined a volume, the pilot's cabin and the rear fuselage another one. In time, the light aircrafts got three volumes:

- the front volume, containing the engine and auxiliary equipment, whose section had the shape of an ellipse with the horizontal long semi axis;
- the central volume, containing the pilot's cabin, whose section had the shape as close as possible to a rectangle with rounded corners;

- the rear volume, which connected the pilot's cabin to the empennage, with a round or elliptical section, with the vertical long semi axis.

The fuselage of large aircrafts is organized in a main volume of constant section (cylindrical), with a frontal part (half of a rotational ellipsoid) and a rear part (mainly conical). The aerodynamic requirements impose the transition between two different sections barrels to be done through a transition surface (blended surface) that will allow a laminar flow. The generation of these surfaces was made by creating a prototype, the projection of curves being made by adjusting the templates manually. The external surface (having a double curvature) had to present continuity in tangency and curvature. It also should not present areas changing form concavity to convexity.

Because of the problems encountered while drawing the fuselage, the method defined remained unchanged till our days and even modern aircraft have the same references, taken from the old naval industry (water line, butt line, stations, port & star side, in/outboard etc.).

Another category of surfaces with a high degree of difficulty are represented by the engine nacelles, air inlets, the wing tips and the door opening frames. If the mentioned surfaces are independent, the transition surfaces between these volumes may have very complex geometry. An example is the wing to fuselage fairing, which is under special aerodynamic constraints, junctions of the trailing edge with the lateral and lower surfaces of the fuselage (Fig. 4.2).

Even in actual projects, which use the powerful Computer Aided Design tools, there are surface areas that still raise problems to the designers. One example is the front - lateral sides of the windscreen and the fin to fuselage fairing.

Fig. 4.2 *The wing to fuselage connection of the DC 3 airplane*

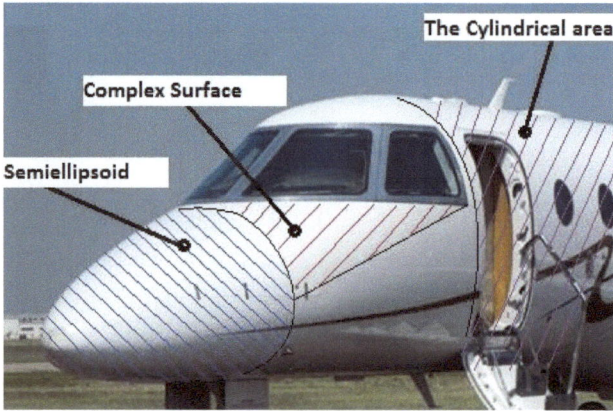

Fig. 4.3 *The connection zone under the windscreen*

In the 90's, sketching the curves was done with the help of elastic rulers with tightly fixed ends; the small connections were being made with the drawing template. The drawing boards could reach up to 6 ÷ 8 meters. A reference for this is the Concorde aircraft, where the wing curves were done directly on a horizontal floor (Fig. 4.4). The ergonomics check of the cockpit was being made with a plastic foil mannequin with joints made with pivots from tap pins.

Fig. 4.4 *The tracing of the curves of the Concorde aircraft [F65]*

The same people that used the drawings with the definition of the surfaces were the ones that made the manufacturing molds (the definition curves), the contour and the flat pattern of the inboard parts that get in contact with the aerodynamic surface, and all outboard parts (for radars and antennae domes, lamps etc.)

The reasons for which the internal structure of aircraft is so complex are the following:

- it supports the aerodynamic surface, with double curvature, with transition areas from edges to surface, etc.;
- for a smooth aerodynamics, all stiffening elements need to be inside the skin. This lead to complex joints, with the need for the transversal and longitudinal elements forming;
- the formed stiffening element (frames, stringers) may initiate the loss of stability, because the fuselage is a thin-walled structure. For of this reason, the sheets forming are standardized.
- the center of the parts are provided with lightening holes and stiffeners for preventing the loss of stability;
- the contour of the parts is minimal, all the flanges being traced around the rivet holes;
- due to overlapping, there are many areas where the sheet metal parts are in packs of three to five pieces, having different orientations and curvatures;

- the sheet metal parts have flanges that are bent to angles varying from 70 to 120 degrees, imposed by the way the web's plane intersects the exterior surface.

This resulted in the structural parts being in large numbers, and attached with a great number of rivets. The complex shape of the surface led to the reduced possibility of using identical parts, resulting in the need of a greater effort in designing and managing the configuration of the structure.

If initially, the complex parts were individually made by tinsmiths, the serial production required the manufacturing of precise molds and forms. The increasing demands of production had led to the need of manufacturing parts that required minimal intervention during assembly ("no shim, no trim"). This has led to an increasing accuracy of the designing process. Parts with complex forms (which had reference to the same surface) and surface joints were calculated with great precision, including the compensation due to material flow in the process of bending after curved edges, with variable angles. This led to high qualified departments, with a large number of technicians and engineers that used tables and complex methodologies for accurate design of the bending curves and flat patterns.

The software CAD programs shyly appeared in the design teams, the first automatized elements being those of making curves of which the coordinates were previously loaded - plotters. Between the first software programs used in aviation Euclid, Unigraphix, Catia, AutoCAD and Solid Works can be mentioned.

One of the most used programs was AutoCAD, which allowed a rudimentary surface modeling. Even so, the engineers from IAR managed to model complex surfaces of some airships (Fig.4.5). These surfaces could not be used for machining or for intersection with planes in order to determinate the framing contours. The aircraft 3D model could be used for any kind of representation necessary for handbooks, product catalogues, etc. AutoCAD offered a very limited possibility for making surfaces with double curvature.

Fig. 4.5 *Cobra AH1 RO and IAR 316b (AutoCAD R13, Eng. Mircea Sasu, 1997)*

Another solution proposed by Autodesk was Mechanical Desktop, which introduced elements of parameterization for the sketch and solid part sections. The surfaces made with this program created a way towards the parametric surfaces used in our days.

Between the dedicated software programs, a special place was won by Catia V5, which made in the early 2000s the generation of surfaces a much easier task. Nowadays, a lot of software solutions present some complex surface generating tools, which create continuities in tangency and curvature radius, convexity and mold extraction analysis. Also, all of them have the possibility of exporting data towards numeric command machines, the manufacturing drawing of the part losing the geometric definition role. The Boeing 787 was the first plane where the problem of abandoning the manufacturing drawings came on the spot.

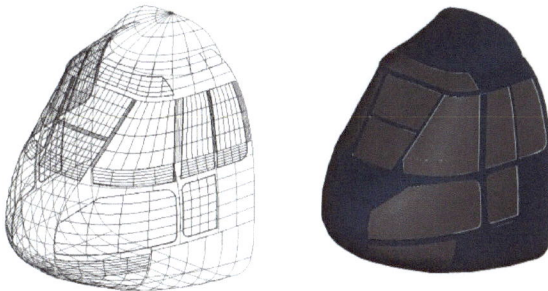

Fig. 4.6 Helicopter *cabin (wireframe and rendering in Mechanical Desktop 6)*

The evolution from software that can only draw 2D projections for plotting purposes to the actual ones was very slow.

Some of the reasons are:

- a completely new philosophy which stood at the base of their elaboration, in comparison with the 2D and 3D nonparametric software;
- the lack of trust in the geometries generated by parametric software from an industry where the curves plotted on blue prints and the plywood templates represented the only standards;
- the lack of a work methodology and experienced trainers;
- working close to the hardware capacity limits;
- very high costs (for proper hardware and – especially – for software).

Therefore, this led to the point where engineers from the design teams were making, through many attempts and a sustained collective effort, their own methodologies, and also having the possibilities to generate the project documents needed for the production and value the capabilities of the software programs.

Because the software programs were very hard to use, a new job has emerged, the engineer specialized in the usage of the computer. By following well based policies, based on feedback from engineers, some software producers set the following as major objectives:

- the software should be facile enough, so that it will allow the designer to focus on project and not to manage with the software limitations
- the software should allow for team work between more designers on a network, respectively a live connection with other departments and services.

It can be considered that these objectives were mostly covered, the aircraft being engineered by multinational teams, located at great distances between each other, working on the same virtual prototype. The current CAD software revolutionized the designing process, the geometry of the parts being precise, the designers being able to correct the deviations, interferences or malfunctions from an early stage. If in the past several prototypes and a big number of pre-serial aircraft were needed to stabilize the production, the actual trend is not to use the prototype anymore.

Regarding the stress calculation, after the pioneering period, the theoretical basics of the flight mechanics and aerodynamics were set. The structural calculation became necessary with the increasing engine power and aircraft capacity. The dramatic nature of the accidents determined the governmental authorities to impose airworthiness regulations, which indicated the in-flight and landing loads the aircraft had to withstand. Due to the complex shapes, the static tests were the only ones to validate the stiffness of the structure for a long period. In time, specialized analytical calculation methods emerged, initially for the statics of the junction areas, followed by the stability of thin wall structure, vibrations and fatigue analysis. After the sustained efforts for developing numeric calculation routines using the computer, methods that allowed the calculation of complex structures were developed. The present is dominated by many software solutions which provide high accuracy simulation of the static and dynamic behavior of structures, for both isotropic and anisotropic materials.

4.1. THE 1903 – 1950 PERIOD

The forces cannot be seen, but their effect can be seen and understood. In the structural analysis the effect of the forces is the one that presents interest.

(F. K. Teichmann)

According to [64] until 1912, the design was chaotic; the basic principles of aircraft design were determined by trials and errors [64]. In [49] the "bizarre" shapes are explained through the fact that many inventors did not posses technical knowledge, that phase of "exploration in aerodynamic science" being followed by the engineering phase.

"The designing of a successful aircraft is not an exact science even in our days" [35]. The history of aviation is filled with airplanes that did not pass the prototype phase; in Romania, before 1945, from 32 new aircraft, only 11 went to series production (the statistic includes aircrafts built by SET, IAR and ICAR).

"The designing of an aircraft involved a combination of scientific proven principles, engineering intuition, market and detailed mission requirements, inventive and a little bit of bravery" [35]. "During the first world war it was mostly intuition, innovation and bravery" [35].

After 1912, an estimation of the performances of the future aircraft, even from the project stage came into light [64]. The design began using a certain amount of rough standardisation and a prototype that did not flew became a rare event. In 1918, ref [50] it is stated that the wing structure became almost uniform for all the aircraft categories.

After 1912, the aircraft design was marked by a scientific approach, based on a design improvement [64]. Still, in [35] it is claimed that the principles of aerodynamics were poorly understood and the structural design was random.

Because of the large number of accidents, in 1913 a committee appointed by the Great Britain Government made the following recommendations:

- the internal wing structure must be strong enough, so it will not fail if the tensile wires fail (for the monoplanes, the tensile wires often failed on landing);
- the doubling (the redundancy) of the main tensile wires and junction fittings;
- attaching the fabric to the ribs and spars so that a crack will not cause the failure of the whole surface.

If in 1914 detailed stress calculations were not used, it had become a common practice after 1918 [35]. It must be mentioned that even in these conditions, a large number of reliable aircrafts were built [35].

In 1914 the first reports of looping and inverted flight and acrobatic figures were recorded, that were added to the load cases the primary structure had to withstand. The development of the fighters required structures that would withstand the load factors encountered in the maneuvers, with the possibility of maintaining integrity in terms of penetration to a certain extent of enemy projectiles, without giving up to the minimum weight [35] (Battlefield Damage Proof Structure).

During the First World War the interest for a scientific approach appeared through the foundation of laboratories – Royal Aeronautical Establishment – Great Britain, research centers in France, Italy, Germany and the National Advisory Committee for Aeronautics (NACA) in the U.S. [35].

In [32] the safety factor is defined (about 1.5), while the military regulations recommended a 10.0 factor for the front fuselage and the engine mount, for the rest of the airplane having the value of 2.5 [Klemin, 1918]. In [50] some calculation examples are given, the wing having a safety factor of 10.0, for the metallic elements values being over 2.0. The safety factor includes [47]:

- the lack of accuracy in estimating the loads;
- the lack of accuracy in the analysis and calculation of the structure;
- the material properties variation;
- the deterioration that occur during the life time;
- variations in the quality of manufacturing.

In 1918 the nondestructive X-Ray control for identifying "latent faults" in metallic parts or for seeing the degree of penetration of the glue in each side of a wooden joint were mentioned.

In [47] it is mentioned that in 1918 the first static test was made for a component of a large aircraft at the Royal Aircraft Establishment. The systematic study of the stress behavior of aircraft structures will be approached after the '50s.

In 1920 it was discovered the lack of resistance of the welded joints under tensile or bending loads [64]. Also in the '20s the aerodynamics and flight mechanics began to develop (process that ended in the '30s) [49]. Starting with 1926, NACA, through Langley Memorial Aeronautical Laboratory, had a decisive role through the publication of the researches in aerodynamics, stability, load calculation and propulsion [35]. The universities had a very important role in research and formation (Massachusetts Institute of Technology, Stanford, New York, Michigan, etc.).

Many research programs have also been financed by the U.S. Army.

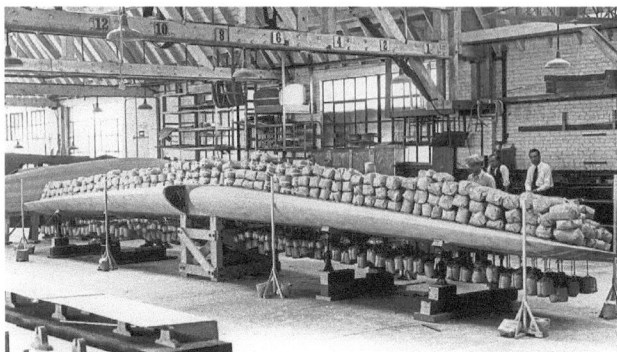

Fig. 4.7 *The static test for the wing of the Lockheed Vega – 1929 [F66]*

If at the beginning of the '20s the layout was imposed by the strenght and stiffness requirements, towards the end of that period the accent fell on the aerodynamic shape [49].

After the latticed beam cease to be the main solution, there were adopted genres specific to machinery components [49].

Until the '30s the airplanes were exclusively statically dimensioned and calculated [35]. The wing was considered an embedded beam [49], the calculation hypothesis being that spar and stringers withstand to the bending, while the skin withstand to the torsion.

Starting with the '30s, the stability calculation started to be approached [35], [27]. Also after the '30s the problem of the wing torsion started to be considered, until then the attention being focused only to the bending [12]. For variable loads designing rules were developed, to prevent the cracks occurence in the structures subjected to fatigue loads [27]. In [27] it is mentioned that, before the Second World War, aircraft had such a short operational period that the number of the stress cycles did not reach significant values.

Fig. 4.8. *The loss of stability of a fuselage [F67]*

In Smithsonian's Institution's "Airspace Mag" it is claimed that the Douglas DC1 is the first airplane built in a scientific way.

In the 40's, the concerns regarding the structure fatigue began. In 1943, there were provisioned such problems due to:

- the continuous increase of the cruise speed;
- the continuation of the primary structure weight saving efforts;
- the increase of the wing load and the maximum take off mass;
- the increase in the operational life of the aircraft, requested by the need of development and production costs amortization.

In 1943 the first design recommendations are set up based on the estimation of the operational lifetime, and the calculation of the flight loads based on statistic data recorded from aircraft with similar speed, mass, dimensions and missions [47].

4.2. THE PERIOD AFTER 1950

Until the '50s the airplanes were designed according to the Safe Life principle. That assumed the ensuring of a well behavior for a part for a certain number of flight hours (after which it is replaced) or for the entire operational lifetime of the aircraft. The testing was done through the static tests, the wing being subjected to a higher load than estimated during flight.

The aviation is a domain where every accident is carefully analyzed in order to find out if there is a lesson that can be learned behind it. The government authorities (especially the Federal Airworthiness Authority) are well-known through the resources invested in finding out the causes, but also through the preventing measures that were integrated in the certification regulations, maintenance programs or operating procedures. Recent aviation history mentions four accidents as milestones for the primary structure integrity after the years of war. These led to the rethinking of the calculation methodologies and the certification requirements. Below the mentioned accidents are shortly presented, together with the lessons learned and the subsequent improvements of the certification requirements [66].

The De Havilland Comet passenger aircraft (1954)

- Three aircraft disintegrate in flight;
- For the first time an investigation that involved great resources was conducted;
- The significance of pressurization over the detail design is understood;
- The influence of the material's fatigue over the safety of the aircraft is understood;
- It was found that because of the limitations in stress calculations, cracks may occur much earlier then expected. Therefore Safe Life would have meant a very short operational period, making the aircraft unattractive from the commercial point of view;
- The Fail Safe principle is introduced – the doubling of the load paths (FAA Regulations, Part 25, Section 1309 mentions: "the isolation of systems, components and elements so that the failure of one does not cause the failure of others);

- The result was a structure easy to inspect and able to withstand loads in case of the occurence of easily noticeable defects. Multiple load paths led to the creation of redundant structures, being able to withstand even after the failiure of one of its members;
- The first tests that subjected the entire aircraft to stress begun (full scale testing – Fig. 4.9.). In the figures 4.10 and 4.11, the conformity between the experimental tests and the wreck of the aircraft may be observed.

Fig. 4.9 *Commet pressurization tests in a water tank [F68]*

Fig. 4.10. *The fuselage failure at the emergency exit – tests [F68]*

Fig. 4.11 *The fuselage failure at the emergency exit – wreck [F68]*

The F111 military airplane (1969)

- The aircraft lost it's wing after a very low number of flight hours;
- The risk presented by the manufacturing damages is understood;
- The Damage Tolerance principle is introduced, which includes manufacturing faults and operational risks;
- The structure is divided in inspectable areas (which must be Fail Safe) and uninspectable areas (which must be Damage Tolerant for military aircraft and Safe Life for civil aircraft).

The B707 cargo airplane

- The aircraft loses it's stabilizer, even it was a Fail Safe design;
- The failure is due to unproperly made inspections;
- The importance of a periodic structural inspection program to be complementar to the Fail Safe principle is understood;
- The structural inspections must be customized for the aircraft that exceed 70% of the operational lifetime ("Geriatric Aircraft");

- The structural inspection must be considered during the designing stage (the concept of inspectability is introduced).

The B737 pasenger airplane (1988)

- The airplane remained without the superior half of the frontal fuselage;
- The combined risk of corrosion and fatigue is understood (stress corrosion);
- The Widespread Fatigue Damage effect, in which a structure can fail through the simultaneous spreading of a large number of cracks, is understood;
- The structural inspection is scheduled not only after a large number of flight hours, but also calendaristhic (in practice it is looking for their sincronisation);

Between other accidents caused by structural failiures, the following are mentioned [lessonsearned.faa.gov]:

The Turk Hava Company, Flight 981, Douglas DC-10 (1974)

The inadequate locking of the back cargo door led to it's separation in flight (Fig. 4.12). This led to the sudden decompression of the cargo compartment and the failure of the passenger compartment's floor, followed by the absorbtion of two seats of three passengers each. The stabilizer's and horizontal rudder's controls were badly damaged, leading to the loss of the aircraft and it's 346 passengers.

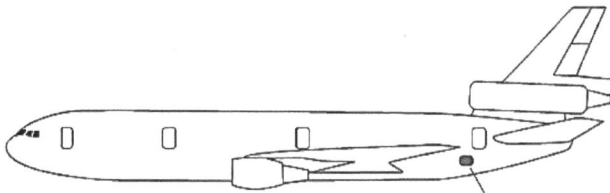

Fig. 4.12 *The positioning of the rear cargo door of the DC 10 airplane* [F68]

Fig. 4.13 *The damaged rear cargo door of the DC 10 airplane belonging to American Airlines [F68]*

A similar accident was registered two years before, on an identical aircraft, belonging to American Airlines (Fig. 4.13). In the Turk Hava Case, the failure of the floor was caused by the maximum loading of the floor. The American Airline flight had only 67 pasengers on board, which led only to the damaging of the flight controls and not to their destruction (Fig. 4.13).

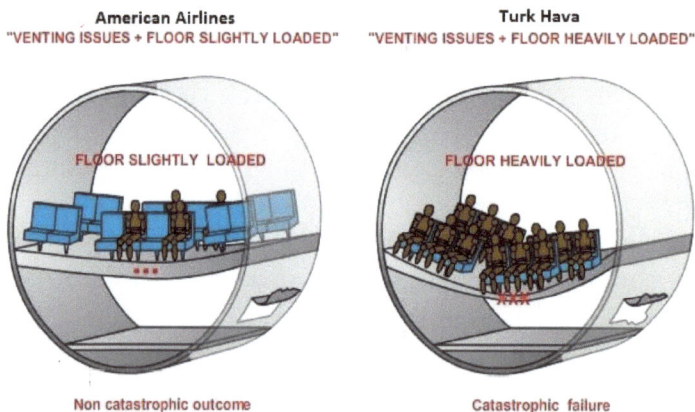

Fig 4.14 *The failure of the passenger floor of DC 10 in the two different accidents [F68]*

American Airlines flight 191, Douglas DC 10 (1979)

Due to the inadequate maintenance, the rear junction fitting of the engine support had failed. The detachment of the engine with the engine support led to the detachment of a portion of the trailing edge of the wing, respectively the loss of electrical power and hydraulic pressure. The crew members and the 271 passengers lost their lives.

Japan airlines flight 123, Boeing 747-SR100 (1985)

Because of an inadequate repair of the rear bulkhead (Fig. 4.15), it failed leading to the decompression of the cockpit and the pressurizing of the bay behind the bulkhead. This led to the failure of the rear APU (Auxiliary Power Unit) firewall which triggered the failure of the structure that sustains the junction fittings of the fin. The failure has led to the loss of hydraulic power and in the end to the loss of the aircraft and its 505 passengers. There were only 4 survivors.

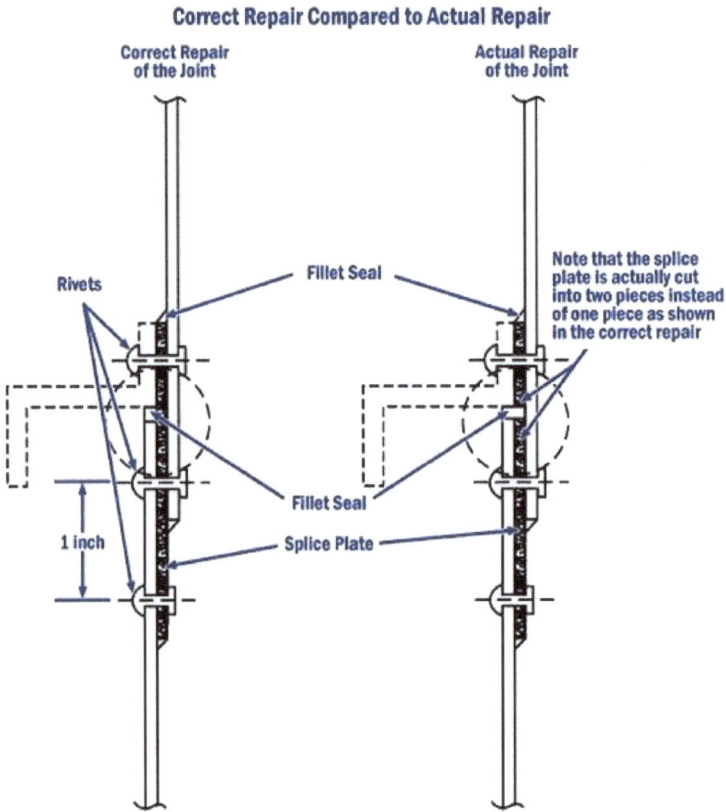

Fig. 4.15 *Detail with the repairing of the rear bulkhead of the Japan Airlines Boeing 747* [F68]

United Airlines flight 811, Boeing 747 (1989)

Due to a failure of the system that safeguarded the locking mechanism of the cargo compartment, it had separated during flight. The decompression of the cargo compartment had led to partial destruction of the fuselage (Fig. 4.17). Nine passengers were sucked through the created hole, the pilots managing to land without any other damage or loss of lives.

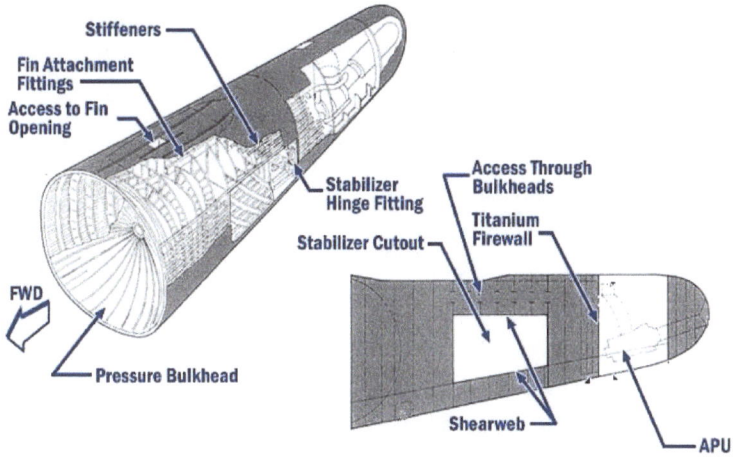

Fig. 4.16 *The structure of the rear fuselage of the Boeing 747* [F68]

Fig 4.17 *United Airlines 811 – the damaged cargo door* [F68]

Continental Express flight 2574, Embraer EMB 120 RT (1991)

The absence of 47 fastening screws on the upper side of the trailing edge of the left part of the stabilizer resulted in its separation, leading to the loss of longitudinal balance and its crash. All the 11 passengers and the crew lost their lives, the cause being a maintenance error.

El Al flight 1892, Boeing 747-200 (1992)

The failure of the bolt attachment of the engine support led to the separation of engine No. 3. In the trajectory after the separation, it hit engine No. 4 leading to its separation and the loss of a large section of the wing. Due to the asymmetry of lift, the aircraft crashed, the crew and 43 people on the ground lost their lives. The accident has led to the improvement of the area of the engine support as seen below:

- the addition of side fittings for engine attachment;
- the replacement of the aluminum fitting attached to the median strut with a titanium fitting;
- the addition of an extra stiffening structure on the wing box;
- the redesign of the bolts attaching the engine support;
- the development of a structure monitoring program through inspections of the engine support after each crash landing or flight in turbulent atmosphere;
- the implementation of improved protection methods against corrosion.

American Airlines flight 587, Airbus A300-600 (2001)

By flying in the draft (turbulent atmosphere) of another passenger airplane, the pilot improperly used the pedal of the rudder control thus bringing the airplane in a yaw maneuver (at a speed beyond the design maneuvering speed). The result was the failure of the junction fittings of the fin (Fig. 4.18) and its separation together with the rudder. Due to the evolution at an angle of 31° form the longitudinal axis, the aircraft lost its engines and, in the end crashed. All the 265 passengers and crew lost their lives.

Fig. 4.18 *American airlines 587 – The fin attachment fittings* [F68]

It is noticed that the failure of the structure or of a fastener can lead to:

- the loss of the aircraft's integrity;
- the loss of control of the aircraft as an effect of the loss of one of the control surfaces;
- the loss of the aircraft's control due to the damaging of the systems routed in the area affected by the structural failure, like the flight controls, hydraulic, electric, fuel, etc.

Nowadays, the certification process requires physical strength tests for material samples, parts and main components. According to [Paul, 2002], the composite structures require a more extended program at static loads and certification than metal structures, while the fatigue is not a problem.

After reference [49], *"The classic calculation methods were based on the strength of materials, by approximating the sizing elements with simple elements (bars, plates, tubes), for each case simplified hypotheses being elaborated and establishing specific calculation techniques."* The '60s space programs have brought the first software programs that could calculate complex structures.

In 1956, Turner introduced the concept of finite element, defining a turning point in the evolution of structural calculation [47]. In the '60s Douglas Aircraft and Bell Aerospace sponsored the development of the MAGIC and FORMAT programs. These were later integrated in NASTRAN (1968) developed by NASA,

Computer Sciences Corporation and McNeal Schwendler Corporation [47], the most longevive and wide-spread structural calculation program in the aerospace industry. In order to reduce the costs and time required for the development, starting in the '70s various manufacturers, private developers, universities or research institutes developed software programs for calculation, design or surface generation. In [62] it is presented a list of over 50 such programs; the great majority being designated to static, stability calculations and structural optimization. The effect of these programs was not only linked to resources, but also to the increase in calculation accuracy, especially for complex structures.

In 1960 Lucien Schimdt proposed the coupling of the finite element analysis with mathematical programming methods for structural optimization. Being defined as a problem of maximizing or minimizing a function objectively supposed to a set of constraints (behavior or manufacturing), it assumes:

- defining of the objective function;
- defining of the constraints set;
- identifying the function variables.

In the '80s the concept of Multidisciplinary Design Optimization – MDO – appeared, targeting constraints and / or functions objectively form more domains (static, aero-elasticity) [47]. Integrated results for the efficiency of command surfaces flutter behavior, tension and movement level and Eigen values were obtained. Well known solutions in this area are Ansys, Nastran, Abaqus, Optistruct.

At the beginning of the '90s, the CAD programs begun to spread. Boeing 777 (1994) was the first plane ever to be designed on a computer.

If initially only the great aircraft producers owned finite element analysis programs, nowadays there are private companies that develop these software programs.

Nowadays the most used programs in structure design are CADDS 5, Catia V4/ V5 and Unigraphix NX. For simulation and finite elements analysis the most wide-spread is MSC Nastran, followed by Ansys, Abaqus, Hypermesh etc.

The finite elements calculation methods are numerical approximation methods, being used as a complement for the classic analytical calculation (hand calculation) in the case of complex structures. In reference [49] it is issued the hypothesis that in a few decades the classic calculation will become "a chapter of history".

Starting with 2000, the concept of Concurrent Design has been put in practice; based on a 3D model (digital mockup), it allows the real time development of structure and installations design and calculation within remote teams. In reference [47] there are presented some examples of aircraft programs requesting 50 to 100 engineering teams, with one exception, in case of F18 Hornet, which required 400 teams (!).

4.3. INTERFERENCES WITH OTHER INDUSTRIES AND SCIENCES

The aviation owes its first progress to the collective effort of inventors and experts form adjacent domains. Otto Lilienthal was a mechanic engineer, Samuel Langley was a well know physicist and astronomer, the adviser of the Wright brothers, Octave Chanute was a successful civil engineer, with experience in railways and bridges (the robust shapes of the biplanes were inspired from the construction of bridges [49]). The chief engineer of the Fokker plants, Robert Platz, was a welding expert [02].

The monocoque structure was applied the first time in aviation by a naval engineer, the similarity between the first monocoque airplanes and ship construction being obvious. The working principle of flight controls and propeller are similar to ships. Also, the lift of a fabric surface was known by the sailors of the ships with side wind navigation. In the Anglo-Saxon teams, the terms of port/ starboard for the aircraft sides, inboard/ outboard, water line for the $Z = 0$ elevation reference datum are used till our days. Also the coordinate system and the names of rotations after the coordinate system axis come from the navy.

The interference with the automotive industry took place in the field of materials, metal manufacturing technology, body structure, drive train and interior design. Form the automotive industry, which is more dynamic, the software programs and efficient work methodologies for designing, stress calculation, production planning was imported. These include efficient surface generation, meshing, boundary conditions, post processing, file transfer between different software etc.

After emerging from the heart of aviation, the space industry exported back to aviation knowledge in the domain of designing extra light structures, the on-board computers (for stability, navigation, etc.) [49] or vectorial thrust.

Most authors consider that the manufacturing of a structure that will comply with reasonable design requirements can be done only through an iterative process, which is also a multidisciplinary process [12]. These iterations (design loops) are the result of the collaboration between specialists in flight operation, aerodynamic, structure, engine, materials, maintenance, onboard systems, manufacturing and quality. Everything has to be under the requirements of the customer, in the end the product having to pass through the long certification process.

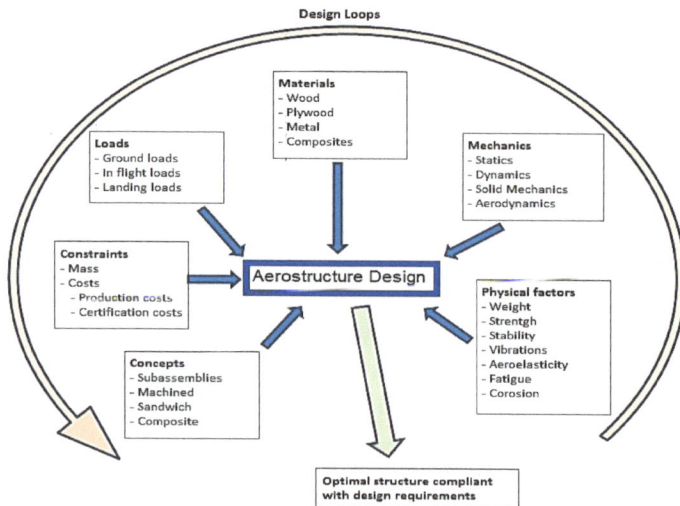

Fig. 4.19 The *sciences contributing to the structural design* [12]

It must be mentioned that the development of the structure (cell) of the aircraft is so laborious that, from a generation to another, it is preferred to equip the aircrafts with new, state of the art engines and systems, while the structure suffers only little improvements. In the operational period, an airplane also suffers, changes of designation on the same structure, an aircraft usually has 4 reconversions (i.e. passengers transport/ medical evacuation/ cargo/ mail). The multifunctional aircraft (which can be used in different missions, i.e. coast guard, border control, evacuation, firefighter) was also requested by the very high development costs. The development of an aircraft comes with big financial risks, a consequence being the small number of successful aircraft (the first one was built only in 1936 (DC-3), after 3 years from the first successful flight).

In turn, aviation generated an important number of sciences, like: aerodynamics, flight mechanics, stability and dynamics. To these it adds up the thin walled structures and aero elasticity, but also significant contributions to the reliability of the systems, risk analysis, quality management and project management. Other sciences tightly related to aviation are medicine, team psychology and the human factors.

It can be concluded that, emerging as a trans-disciplinary domain, the aviation became a multi-disciplinary domain, but well outlined, and in time contributing to the progress of other industries, raising new safety and quality standards, which in turn led to the birth of new sciences.

CHAPTER 5

Structural joints

The structural joints represent the area where different elements are rigid or removable joint. The joints connect different aerostructures (for instance: empennage with fuselage, the landing gear with fuselage) or between elements of the same aerostructure (for example: stringers to frames, longeron to columns and bracings). If a structure withstands the aerodynamic and payload's inertial loads with a volumetric structure, presenting a large inertia momentum, a joint takes over only concentrated loads having a very limited volume. For this reason, the aerostructures are thin-walled, from lightweight materials, while the joints are massive parts made by steel, titanium or duralumin.

The big difference between the values of the loads present in joints and aerostructures respectively, requires a transition area, to dissipate the concentrated efforts from joints to the thin walls of the aerospace structures. A sudden change of the section area leads to big stress concentration factors and the local failure of the thinner wall exactly near to the solid wall of the joint. The structure design was always a difficult process, because the designer has to realize a lightweight structure but robust enough to resist the loads and stiff enough to not alter the aircraft aerodynamic behavior. Investigating how the joints have evolved over time, a big variety of materials and assembly procedures may be observed, reflecting a long process of searching for solutions to eliminate the shortcomings of previous joints.

5.1. NOTES ON THE EVOLUTION OF STRUCTURAL JOINTS

The latticed beam was the first structure able to provide a feasible compromise between the strength, stiffness and available volume requirements, under the condition of minimum structural weight. For this reason, the lightweight latticed beam emerged with the first controllable aircraft (1903) and it is still used on a large scale at the current aircrafts.

A specific aspect of the aircrafts is the complexity of the structural nodes; due to the aerodynamic shape of the external surface, the supporting structure is complex too; thus, the rectangular parallelepiped or the prism shape, normally used at the beam-type structures, have been transformed into truncated pyramid, in plan, the rectangle being replaced with a trapezoid or an irregular polygon (especially at helicopters). It can be stated that the only "rule" is the symmetry.

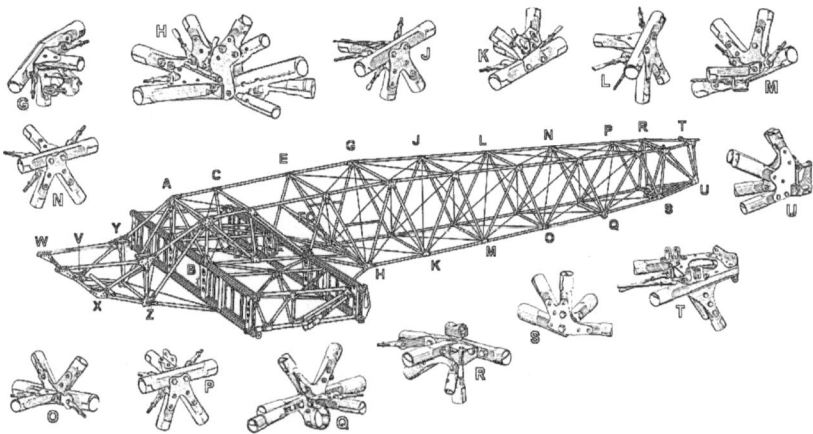

Fig. 5.1 *The fuselage nodes (Hawker Hurricane, 1935) [F69]*

The replacement of the tensile wires with bracings led to the realization of constructions stiffened in all planes, but also to very complex nodes. In the joining points of the fuselage (wing, empennage, landing gear, engines) the structural nodes have been additionally stiffened to withstand to the concentrated efforts. It observes a very wide variety of solutions, from one manufacturer to another, and even from one aircraft to another - it can be stated that there is no uniformity of the design and, of course a lack of standardization. Moreover, there are numerous examples of aircraft where each node geometry or even the conception is different (Fig. 5.1).

Before 1903, when it was considered that the bat wing and ornithopters were the ideal solution for flight, the main elements were the wood or bamboo slender sticks, the metal wire and the textile rope. The wood sticks were embedded in the sewn fabric on the perimeter. For the wire attachment to the end of wooden sticks metallic clips or thimbles by sewn leather were used.

The first wooden aircrafts employed, from the beginning, stiffening elements of joints as corner brackets or hollow clips, bolted to the wooden structure. Originally, made of Aluminum, after 1915 these were manufactured from hot or cold formed steel [54].

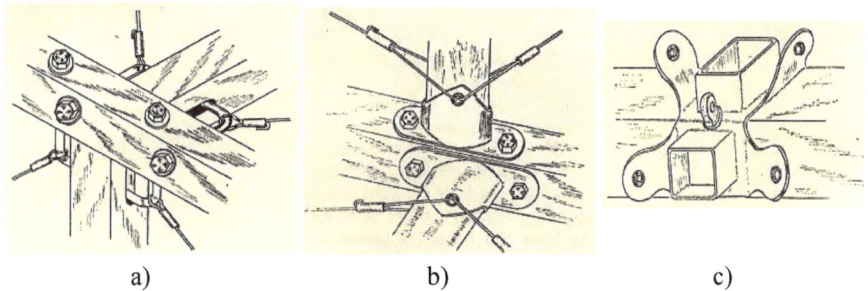

a)   b)   c)

Fig. 5.2 *Joint brackets: a) threaded rods; b) welded Aluminum; c) formed and welded [54]*

The most representative flanges (Fig. 5.2) were threaded rods of "U" shaped (Bleriot), Aluminum welded flanges (Deperdussin) or formed sheet metal parts with welded areas of pillars insertion (Aviatik). The flanges for one joint contained also the eyelets for tensile wires attachment.

Due to the problems associated with the wood and fabric, the producers sought solutions to increase the durability, the impact and fire resistance, and to eliminate the anisotropy and the network of tensile wires.

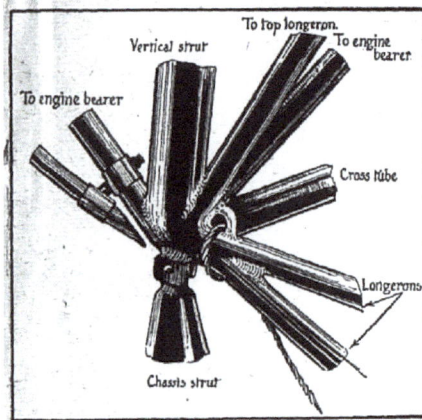

Fig. 5.3 *Structural node with 8 members used by Fokker [F21]*

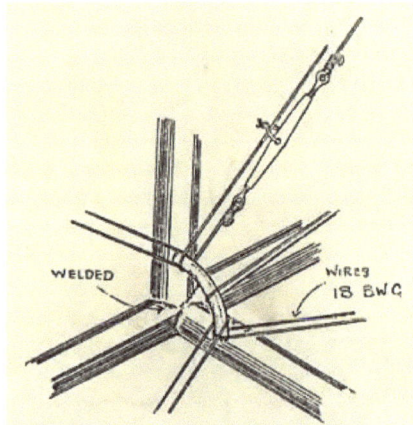

Fig. 5.4 *Structural node with tubular gusset [08]*

With the transition from wood and bamboo to metallic tubes, the need for a new kind if joint appeared. As early as 1907, Fokker used on its aircrafts the welded structure (Fig. 3.2). In

1918, the Flight magazine remarked for one of Fokker biplanes that the structural nodes used are the result of the excellent skills of the welders.

In 1919, Camm considered that the metallic tube is the most practical shape to be used within aircraft structures. In figure 5.4 a detail of the joint with tubular curved gusset, having the role of support for the end of the tensile wires is shown. Also in reference [8], the use of the spruce wood inserted into the tube to prevent the buckling is mentioned.

5.2. STRUCTURAL JOINTS WITH ASSEMBLY ELEMENTS

The technological difficulties due to the insufficient knowledge of the welding process and the weld seams failure due to fatigue led to the search for solutions to replace the welded joints; in certain situations, the joint type asks for the replacement of the tube with riveted or opened profiles.

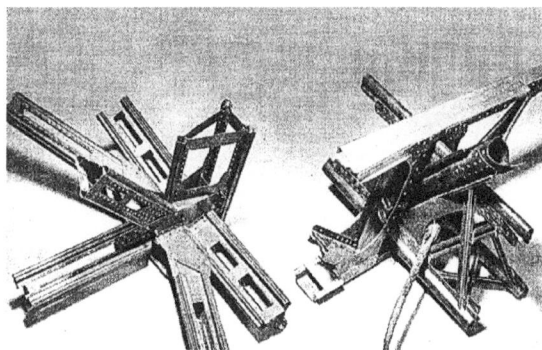

Fig. 5.5 *Structural node of the Dornier Rs. I flying boat [73]*

The riveted structure started to be used in aviation very early, being introduced by the engineers from Zeppelin. Having the experience of the large airships structures, they have brought in aviation, the lightweight latticed beams, having structural details specific to the lightweight structures. By applying this knowledge to seaplanes (flying boat) - the largest aircrafts of that time – the aerospace structure became a latticed beam made by hollow structures, which has led to extremely complex nodes (Fig. 5.5).

After the unsuccessful attempt of the realization of the wing with integral metallic structure within the J1 airplane, Junkers was focused on the multi-planar beam structures. These structures used special nodes that connected both closed profile members and opened profile members (rarely used in aviation) (Fig. 5.6). The joints were reinforced with steel drawn riveted flanges (Fig. 5.7).

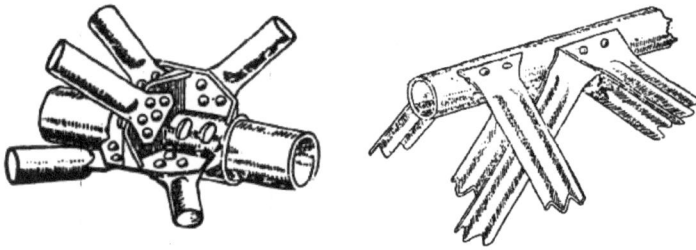

Fig. 5.6 *Structural nodes of Junkers wing [F70]*

Fig. 5.7 *Joint of different profiles (Junkers, 1924) [F12]*

Other design solutions have used riveted milled tube ends. The welding was replaced because the circular weld seams were subjected to tensile loads, while only shear load is recommended. In figure 5.8 a) it is shown an example of tubes with milled ends, with joints of lug – fork type, bolted together. In figure 5.8 b), a flat bracket with tubes having the ends assembled with tubular rivets is presented.

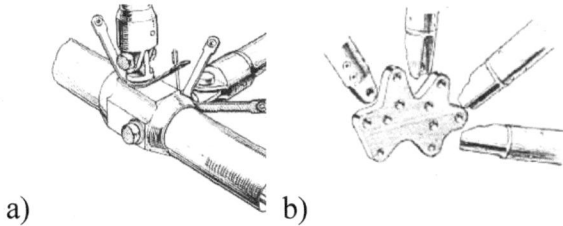

a)   b)

Fig. 5.8 *a) Joints with milled ends (Sidestrand, 1926); b) Tubes with riveted end forks [F21]*

In 1928, Blackburn Lincock used a hybrid solution by riveting the sheet metal brackets to the end of the beam members. The brackets were attached with tubular rivets (Fig. 5.9) that over time did not provide satisfactory results; currently, this type of rivets are not used anymore in structural applications, their use being limited to flight controls. It may be observed that the columns and bracings are riveted, their subassembly being attached with bolts to the longerons.

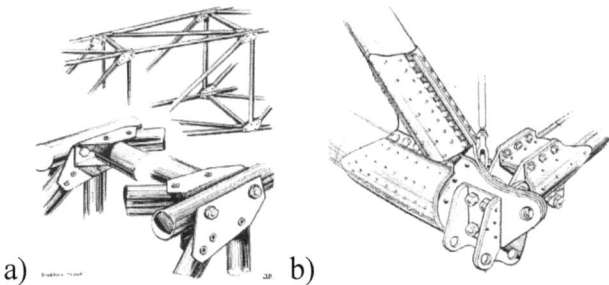

a)   b)

Fig. 5.9 *a) End flanges, riveted to tubes (Blackburn, 1928) [F24]; b) Tubes with riveted collars (Shorts Valetta, 1930) [F21]*

The Shorts Company has found another solution to assure a smooth transfer of the stresses from the structure to the joints by inserting gussets in the symmetry plane of the members. The connection between members and gussets was made by symmetrical doublers (collars), riveted with blind rivets (Fig. 5.9

b). Although this concept was very robust and it allowed a longer period in service, the labor costs were very high. Note that for the modern aircrafts, the blind rivets are not accepted for the primary structure.

The lightweight structures use thin walls; therefore the buckling is the most common failure mode. In order to prevent the buckling occurrence, in 1929 Bristol used the wavy structure members made by corrugated sheet metal. In the Flight magazine (1929) it is mentioned that the pillars made by corrugated sheet metal have a higher buckling strength than those of circular section (Fig. 5.10 a). After an assessment elaborated by the author, the following conclusions were drawn:

- for columns with the same wall thickness and equivalent critical general buckling force, the weight of the column with corrugated walls is with 40% higher, for a critical local buckling strength of 90% of the straight tube;
- for columns with identical weight and wall thickness, the critical general buckling force of the corrugated tube is reduced with 35%, the one of the general buckling remaining of 90% vs. the straight tube;
- the corrugated columns don't bring any gain regarding the weight or the buckling behavior by geometric shape; the only benefit can be recorded by the local hardening of the sheet metal after bending, leading to superior mechanical properties.

After 1930, in aviation, the spot welding started to be used. In figure 5.10 c) an example of the Stal 2 airplane with a longeron having the flanges with a laborious and expensive structure is presented. In this case, the design approach was closer to architecture than mechanical engineering. Currently, the spot welding has very limited applications on the primary structure of the aircrafts.

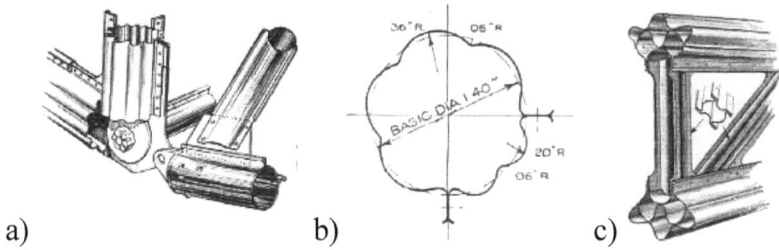

a)   b)   c)

Fig. 5.10 *a) Corrugated sheet members (Bristol 110A, 1929); b) Corrugated sheet column section; c) Corrugated sheet longeron (Stal 2, 1931) [F21]*

In 1934, Shorts used the structure members from riveted profiles in order to improve the buckling strength. The gussets were used not only to connect the structure members, exceeding the joint area. The solution reminds to the beams of the metallic bridges (Fig. 5.11). Members with riveted profiles have been used by many airplanes, for longerons or high loaded stringers. For the closed section members (hollow structures) the omega or *U* profiles are common, the solution being further used also for composites.

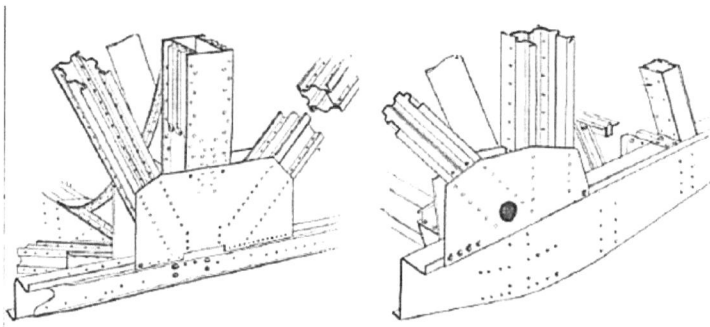

Fig. 5.11 *Members with riveted profiles and gussets (Scylla 1934) [F24]*

In 1928, the magazine Flight presented a structure with pillars made from riveted omega profiles, with gussets inserted in the symmetry plane, which was distinguished by simplicity, even if it presents a large number of rivets (Fig. 5.12). This system was developed to improve the shock behavior (hard landings) [Flight Feb. 1928]. The inserted gussets dissipate the stress peaks from nodes and reduce the buckling length - a 20% reduction of the buckling length leads to 50% increase of the buckling critical force.

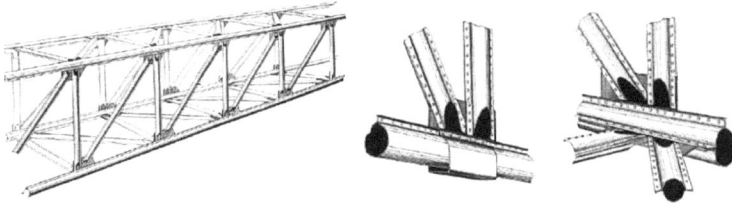

Fig. 5.12 *The beam with riveted omega profiles [F24]*

5.3. STRUCTURAL WELDED JOINTS

In parallel with the development of new concepts, different solutions to improve the structural nodes were tested. Between the encountered solutions, the outer collars (Fig. 5 .13 a) and the welding of flanges inserted into the tubes ends (Fig. 5.13 b) are distinguished.

The solutions with collars and doublers are still used in civil engineering, in aviation remaining only the simple welded structures with or without gussets (Fig. 5.14).

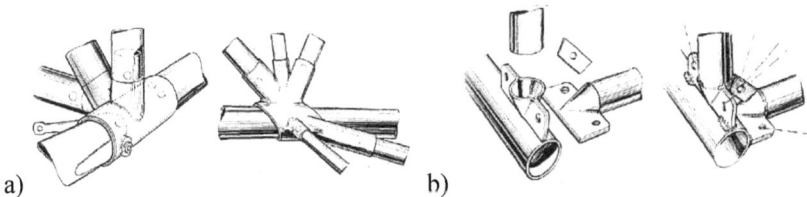

a)   b)

Fig. 5.13 *a) Joints with collars; b) Joints with flanges inserted in tubes [F21]*

In 1930, the Avro Trainer presented several new solutions, which have been preserved and have developed over time; among these there are worth to be mentioned:

- bushings inserted into tubes in order to allow the attachment of outer elements with bolts (Fig. 5.15. a);

- formed end tubes for the joints with tubes of smaller diameters (Fig. 5.15. b);
- the use of gussets tangential placed to the joint, allowing also fittings attachment (Fig. 5.15. c, d).

Although over time there have been many alternatives to the welded joint, in aviation only the welded latticed beams are currently used. After 1940, welding started to be used less on airplane fuselage. This was used extensively on the helicopters' fuselages and tail booms, till the end of the 1970's when the metallic monocoque started to prevail. For airplanes, the welded latticed beams are met only at the fuselage of the light aircrafts.

Fig. 5.14 *Pre-assembled structural node - Heinkel He 51 (replica) [F71]*

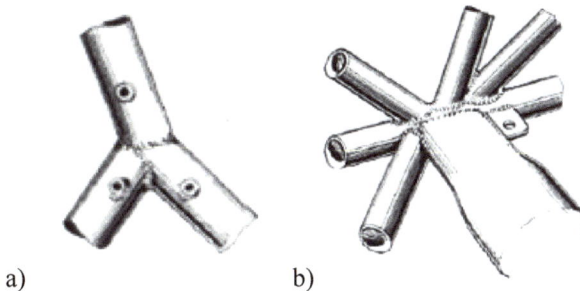

a)   b)

189

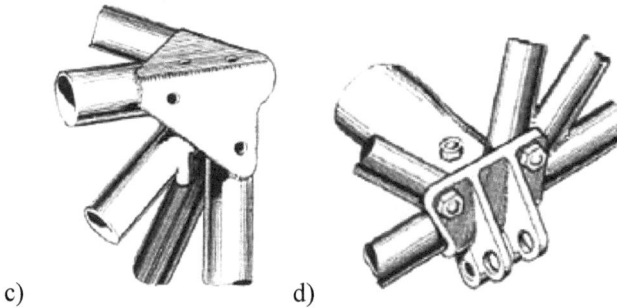

c)   d)

Fig. 5.15 *a) Bushings inserted into tubes; b) Joined tube with formed end; c) Gusset for fitting attachment; d) Fittings attached by the welded joint [F24]*

5.4. MATERIALS USED IN WELDED STRUCTURES

The most used metallic materials are those with low carbon percentage, welding inert gas (WIG) being the most common process. The commonly used steels are molybdenum chromium steels SAE 4130 or 15CVD6. The stainless steels are used only for fasteners and non-structural applications.

The non-metallic materials which can be welded with good results are: 1100 3003, 5052, 4043 or 5086; these are materials used in non-structural applications (5086 is distinguished by a very good plasticity, allowing deep forming); using appropriate thermal treatments 6053 and 6061 can also be welded [71].

The magnesium-based alloys can be welded in good conditions using the WIG process, but precautions related to the after welding deformations must be taken [71].

In most of cases welding join parts made from the same material. If different raw materials are used, technological compatibility has to be previously checked.

For high strength and lightweight applications or for areas exposed to high temperatures, titanium may be used in various alloys; the welding process is similar to steel and aluminum alloys.

190

5.5. TUBULAR WELDED STRUCTURES

The structures using closed section members CHS (Circular Hollow Section) and RHS (Rectangular Hollow Section) are widely used in various types of applications extending beyond the aviation field, as presented in references [34], [45], [65] and [68].

As usual, the tubular structures are met in the civil engineering, offshore platforms and energy industry. Particularly for aviation, these are used either for primary structures, or to withstand the concentrated loads and their transfer to the thin wall structure of the aircraft.

Most of studies are focused on the static, dynamic and fatigue behavior of the joints. The results provide an overview of CHS and RHS with all types of used joints (*T, Y, K, X*) with gap or overlapping bracing (Fig. 5.16).

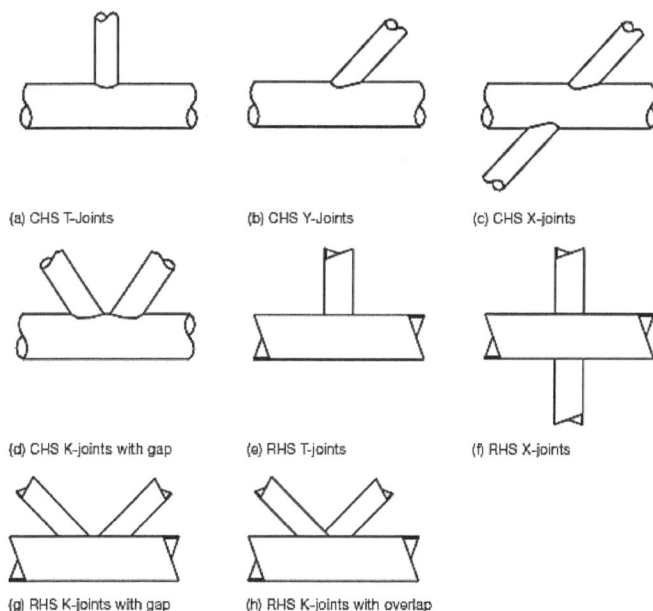

(a) CHS T-Joints   (b) CHS Y-Joints   (c) CHS X-joints

(d) CHS K-joints with gap   (e) RHS T-joints   (f) RHS X-joints

(g) RHS K-joints with gap   (h) RHS K-joints with overlap

Fig. 5.16 *Classification of the planar joints of CHS and RHS [68]*

The surveyed literature deals with planar or spatial joints, under axial load, in plane bending or out of plane bending. Different approaches of the connection cases between tube structures and I or U profiled beams are presented in [34] and [65]. The main reasons for the use of tubular structures vs. laminated (opened) profiles are the good stability behavior and the lightweight properties. There are researches studying the bamboo wood, in order to get the similar optimal solution the nature offers for the need of rigidity under fatigue loading.

In reference [34], it is stated that for the same critical buckling force, a tubular column (CHS or RHS) will only weight 60% of a I (or H) profile. The same reference mentioned that the most common configuration for combined efforts is between same-type members (CHR/CHR or RHS/RHS), other combinations being very rare.

On heavy structures (halls, stadiums, bridges), the gussets are used to facilitate the addition of bracings or other elements attached to the RHS.

Reference [34], offers some design guidance related to limit deformations for the RHS and the distribution of stress in the gussets.

As [36] shows, in civil engineering the gussets are used to join the horizontal beams to columns (seat bracket) specifically connect the columns to the basement plates.

Studies related to the stiffening of the CHS T joints using gussets, chord doubler or the external collar were developed in [39] and [45].

According to [Blodgett, 1976] the gussets are used to decrease the stress level in a K joint, while [39] describes the adding of a base plate (chord doubler) as a cost effective solution to damaged structures.

A study related to the decrease of the maximum stresses in the K joint by adding a gusset was realized by Nazari and Durack [38]. Based on this study, it is demonstrated that the stress concentrator factor (SCF) decreases approximately 45% for the axial loads, 33% for in plane bending and 18% for out-of-plane bending.

The aviation applications traditionally use the circular hollow structures (CHS), the main reason being the minimum aerodynamic drag for an acceptable bending behavior, despite the geometry shape problems raised by the members in the connection area or due to the multiple welds.

In the same time, CHS are widely used for planar or spatial structures, often presenting gussets in different types of joints or in the fittings area in order to decrease the stresses level or to improve the rigidity.

5.6. WELDED STRUCTURES WITH GUSSETS

The gussets are often met in welded structures of the aircrafts. These are added to the joint or they substitute certain tube segments to reduce the number of elements in the structural node (Fig. 5.17).

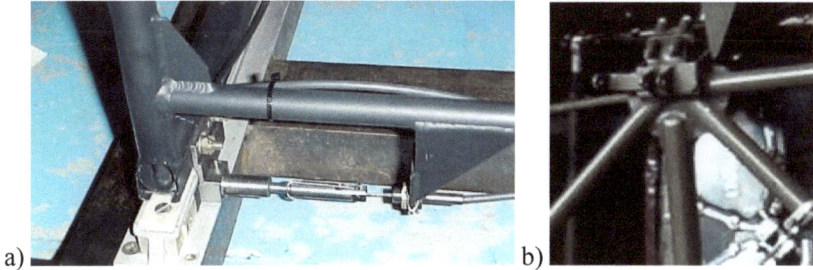

Fig. 5.17 *Details of structures with use gussets: a) passenger seat; b) SA316B fuselage [F18]*

According to reference [43], the tapered gussets should be included in all important welded joints to ensure the gradual stress distribution in the joint's elements; also, the gussets reduce the fatigue risk by reducing the stresses level (Fig. 5.18).

The literature presents numerous examples of gussets. The shape, the proportions and the dimensions varies according to authors (Fig. 5.19). The gussets may be placed radial or tangentially, external or inserted into the members. In the structural nodes with diagonal elements, the gussets can simplify the node geometry and welding.

In reference [20], the gussets are recommended for joint reinforcement, adding strength and stiffness. They are also recommended to increase the safety. In the same reference is mentioned that the gussets added to a structure increase also its torsional stiffness.

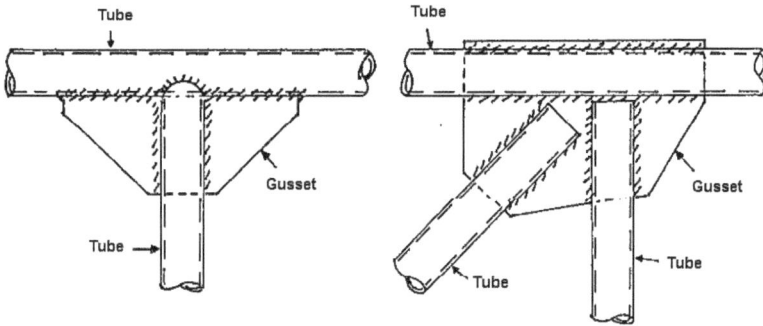

Fig. 5.18 *Gussets placed radially or integrated in joint* [43], [06]

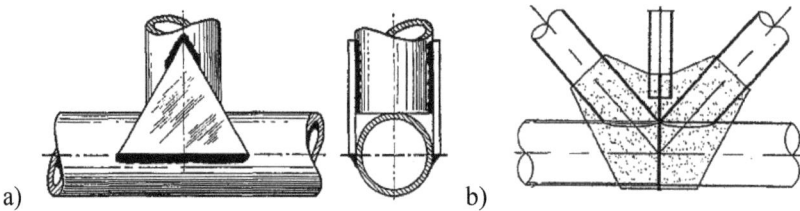

Fig. 5.19 *Examples of gussets: a) triangular shape [46]. b) gusset placed in the symmetry plan [38]*

In [06], gussets are recommended for areas subjected to vibrations or/ and out of plane bending (Fig. 5.20). F. Bruhn describes the use of several types of gussets, but it is not recommending any pre-dimensioning method or the use of a particular type of gusset according to the application.

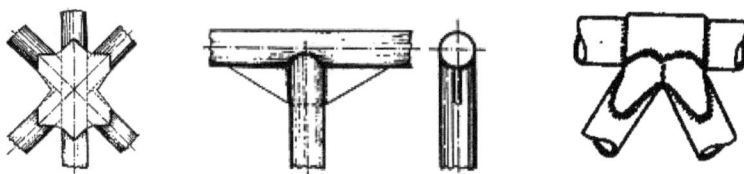

Fig. 5.20 *Gussets for the areas subjected to vibrations [06], [43]*

Reference [24] presents the gussets with profiled edge (to prevent the buckling, and to increase the weld seam width at the gusset end) - (Fig. 5.21 a). The same author presents a spot welded gusset; a particular welding made between the edges of a hole cut into the gusset and tube itself (Fig. 5.21 b).

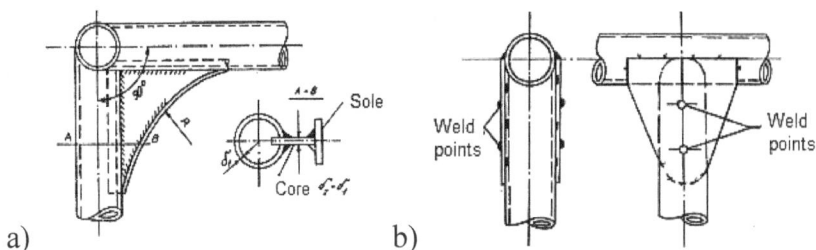

a)   b)

Fig. 5.21 *Gusset with welded free edge; b) spot welded gusset [24]*

In reference [20] is shown the use of the gussets for the following reasons:
- the requirement an additional length of the weld seam (Fig. 5.22);
- the tubes length may be decreased, so that the entire load to be taken over only by the gusset;
- the direct transfer of the force to the primary member is allowed when the elements are placed on both sides.

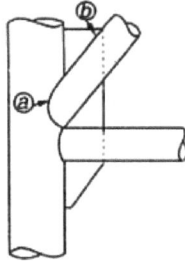

Fig. 5.22 *Welding (a) is secured by welding (b) [20]*

In reference [20] is also raised the problem of the tubes stress level increase at the gussets end due to the increased stiffness of the joint. This effect can be reduced by the curvature of the gusset free edge (Fig. 5.23).

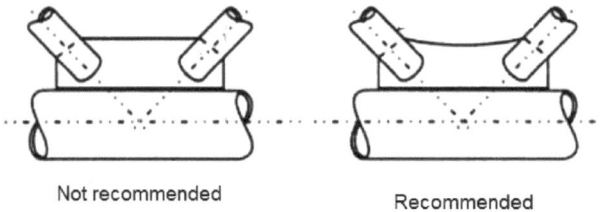

Not recommended   Recommended

Fig. 5.23 *The gusset with curved edge to reduce the stresses in tubes [20]*

In reference [52], the rounded adjacent tubes ends are recommended as a common solution in aviation (Fig. 5.24 a). During the interwar period, the tubular gussets were widely used, especially for the possibility of diagonal tensile wires attachment (Fig. 5.24 b).

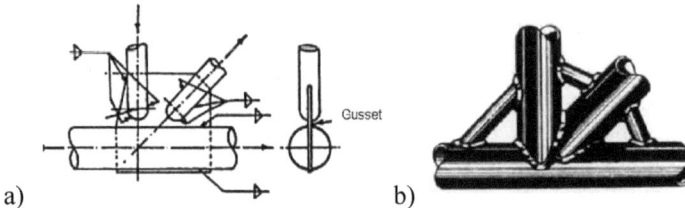

a)   b)

Fig. 5.24 *a) Symmetrically placed gusset, with tubes without contact with the main tube [52] b) Tubular gussets [F72]*

In reference [20], the fundamentals of the gussets placement are presented. The problems raised by the placement of the gusset in the joint symmetry plan are:

- *cracks at the gusset end;*
- *deformations of the main tube.*

It is important to note that there are many welded structures with no use of gussets. This problem is due to the fact that the design approach in the aviation industry is not homogenous, depending on the design team and methodologies.

It can be concluded that there is a great variety related to:

- the flat shape of the gussets (dimensions' ratio, geometrical shape of the free edge);
- dimensions of the gussets relative to the tube's dimensions;
- gussets' placement (radial, inserted into the joint or tangent to it);
- gussets that are replacing the tube joint (the tubes do not come into contact) or just strengthen the joints;
- the gussets number;
- joints with or without gussets;
- rectangular frames, stiffened with gussets vs. frames stiffened with bracings;

Also, a lack of information related to the material thickness may be noticed.

5.7. WELDED STRUCTURES - DESIGN RECOMMENDATIONS

According to references [6], [43] for high quality welds, the designer should use the following best practices:

- weld seam subjected to shear and not to tensile loads;
- components of the same thickness or with the maximum thickness ratio of 2 : 1 (exceptionally 3 : 1);
- for thin walls parts (less 1.0 mm) the welder must control the temperature to prevent overheating (leading to sheets

melting and walls penetration); in [Blodgett, 2005], it is recommended to avoid welded sheets thinner than 0.9 *mm*;

- avoid weld seams placed perpendicular on the tube axis (subjected to tensile loads); these have to be placed at 45 ÷ 60° with respect to the tube axis;
- avoid as much as possible overlapping of weld seams;
- avoid weld seams on both sides of the thin sheets (to prevent the material melting);
- avoid small distance between weld seams (cracks due to the small contraction space may occur);
- avoid weld seams on the bent sheet area (cracks may occur);
- avoid brackets weld in the middle of the structure members (not to affect the buckling behavior); if needed, they will be screw mounted with collars;

5.8. TRENDS IN THE WELDED STRUCTURES DESIGN

The current trends show that the composite materials take over more and more field. The welded structures have become increasingly rare in the primary structures, due to the poor behavior to fatigue, where riveted structures obtained improved results.

The welded joints present the disadvantage of cracks propagation through the whole assembly (the crack propagates through the weld seam). In the case of the riveted structures, the crack and the potential failure is maintained at the level of the affected component, the residual strength of the remaining structure being clearly superior. In this way, it explains why the welded structures are not used anymore for the primary structure of the pressurized aircrafts; this configuration is maintained only at the light aircrafts.

The following trends in welded structures design may be mentioned:

- the replacement of the integrally welded latticed beams with beams with forged nodes or integrally machined latticed beams;
- the replacement of a structure with a single one machined part thanks to the new technologies that have allowed both large machined volumes and shape complexity for significant cost savings;
- the replacement of the flat shapes structures with honeycomb sandwich structures;
- the replacement of metal with composite material structures;
- stringers welding in areas that are not subjected to fatigue and corrosion (for the fuselage panels).

However, there are areas where the welded structures maintain their own niche, the possibility to replace them being quite small:

- prototype and small manufacturing series structures;
- landing gear of the light aircrafts;
- engine support for the piston and turbojet engines;
- equipment racks (filters, tanks, optional installations) which require a complex geometry and large dimensions;
- fuel tanks (not integrated in the wing);
- seats structures;
- cowlings, air inlets, doors margins, exposed to wear or accidental impacts;
- air conditioning ducts.

The researches in welded structures field are traditionally dedicated to various types of welds and joints - establishing and testing technologies and calculations of new design. The current researches are focused to the new technologies and materials, following the innovation strategy emerged from the energy saving needs, environmental friendly aircraft and cost reductions.
Regarding the configuration, there are many studies about the joint type, combination of different materials, manufacturing technology and their influence on:

- static strength;
- fatigue strength;
- residual stresses.

There are also researches related to more accurate calculation methods using the finite element analysis, namely the effective modeling of the weld seams and the adjacent areas.

5.9. COMPARISON BETWEEN THE COMMON STRUCTURAL JOINTS

In Table 3.1 an overview about the welding vs. the other assembling processes used in aviation is presented. The riveted assembly was taken as a reference, being the most used for aerostructures. The pressed assemblies are not considered, having a very limited application, being conditioned by the presence of a shaft/hole fit.

The biggest disadvantage of welding is related to the cracks propagation; for all other assembling methods, the crack stops in the part where it has been initiated; in case of welding, the assembled parts behaves as a whole, the crack passing by the weld seam in all adjacent parts. Another problem is related to the nodes where there are more than three tubes; the weld overlapping is not recommended, because the mechanical properties of the first weld seam may be altered.

Between the disadvantages of the technological welding process, it may be mentioned [29], [43]:

- tooling fixtures for positioning and pre-mount;
- the tubes ends trimming by milling, which the case of multiple nodes is a complex operation;
- the thermal expansion/ contraction of components during the welding process; the deformation of the whole assembly due to the accumulation of residual stresses after the welding. These problems are eliminated in practice by the placement of the weld seams, the welding sequence, the preheating the area to be welded and the post curing.

- mechanical properties changing of the material after welding; according [Niu, 1988] a reduction with 10% of the calculation after welding ultimate stress is recommended;
- occurrence of the welding errors, which decreases the strength of the weld seam;
- high skilled operators;
- expensive special procedures of non-destructive quality control (color penetration, ultrasound or X-rays).

Table 5.1 Comparison between different assembly processes
(0 – reference, + better behavior, - poorer behavior)

Criterion		Welded assemblies	Screw mounting	Riveted assemblies	Bonded assemblies
Mechanical loadings	Static	+ +	+	0	+
	Dynamic	+	+	0	0
	Vibrations	-	-	0	0
	Fatigue	- -	0	0	+
	Impact	+	0	0	-
Safety		-	+	0	-
Corrosion		0	0	0	-
Weight		-	-	0	+
Manufacturing		+	-	0	+ +
Repairs		0	+	0	-
Costs		+	-	0	+ +

As advantages in comparison with other manufacturing methods, in reference [11a] there are mentioned:

- sub-assemblies with more rational shapes;
- a more economic use of the material and small technological adds-on;
- possibility of automation and the reduction of the manufacturing time;
- cheap and easy to maintain welding machines and tooling;
- no fasteners needed;
- superior strength, if the long weld seams are used.

Recent assessments between the riveted, welded and machined structures (wing panels), regarding the fatigue behavior, elaborated by [69] indicates:

- a superior behavior of the welded and milled structures vs. to the riveted ones regarding the stress concentrators occurrence;
- the lowest increase of the crack growth rate of the welded structure, but the smallest residual strength (Fail Safe behavior) after the failure of two stiffening elements (stringers);
- cracks propagation in the welded structure depends very much on the area where these are initiated, the highest propagation rate being of the cracks initiated in the area affected by welding;
- welded and machined structures require skin bonded doublers to delay the cracks occurrence and to decrease their propagation rate; between the titanium and carbon fiber doublers, the last ones proved to be more effective in the loads transfer in the case of a cracked structure.

References

[01] ALEXANDRU, St.; COTTA, N. *Utilajul şi tehnologia fabricării produselor industriale din lemn.* Bucureşti: Editura Didactică şi Pedagogică, 1964.

[02] ANDERSON, J. D. *Introduction to flight.* McGraw Hill, 2004.

[03] ANTONIU, D. ; CICOS, G. *Romanian Aeronautical Constructions.* AACR, 2003.a

[04] BALOTESCU, N., s.a. *Istoria aviaţiei române.* Bucureşti: Editura ştiinţifică şi enciclopedică, 1984.

[05] BIDDLE, W. *Barons of the Sky: From Early Flight to Strategic Warfare.* Johns Hopkins University Press, 2001.

[06] BRUHN, E. F. *Analysis and design of flight vehicle structures.* Tri-State Offset Company, 1973.

[07] BUCHANAN, S. *Development of a Wingbox Rib for a Passenger Jet Aircraft using Design Optimization and Constrained to Traditional Design and Manufacture Requirements.* CAE Technology Conference/ Altair Engineering, 2007.

[08] CAMM, S. *Aeroplane construction.* London: Crosby Lockwood, 1919.

[09] CAUGHEY, D. A. *Aeronautical history. Important advances în aircraft design.* New York: Cornell University, 2008.

[10] CERVELLERA, P. *Optimising Driving Design Process: Practical Experience on Structural Components.* Bari: Proc. 14th Convegno Nazionale ADM/AIAS, 2004.

[11] COLLARD, D. *Concorde Airframe Design and Development.* Zurich: Swiss Association for Aeronautical Sciences, 1999.

[11a] CONSTANTIN, E. T. *Proiectarea masinulor, utilajelor şi construcţiilor sudate, Suport de curs.* Galaţi: Universitatea din Galaţi, 1981.

[12] CORONA, E. *Notes on Aerospace Structures.* AME 30 341, University of Notre Dame, 2006.

[13] CUTLER, J. *Understanding Aircraft Structures.* Wiley, 1999.

[14] DICAIRE P. et al. Isogrid CNRC-IRAP. Montreal: Aerospace Innovation Forum, 2011.

[15] DIMA, G.; BALCU, I. *Consideraţii privind proiectarea pentru masa minimă a aerostructurilor.* In: Recent, Vol. 13 (2012), No. 3 (36), pp. 259-266.

[16] DIMA, G.; BALCU, I. *Tendinţe actuale ale Light Weight Design (LWD) pentru aerostructuri. Interferenţe cu industria auto.* In: Buletinul AGIR, Nr. 1/ 2014, pp. 58-63.

[17] DIMA, G.; BALCU, I. *Notes on Aircraft Structures Welded Latticed Beams Joints.* In: COMEC Proceedings, Brasov, 24-25 oct., 2013, pp. 119-125.

[18] DIMA, G.; BALCU, I. *Actual Status of Gusseted Joints of Welded Structures form Aerospace.* In: COMEC Proceedings, Brasov, 2013, pp. 112-118.

[19] DIMA, G.; MACHEDON–PISU, T.; BALCU, I. *Alternate Design Solutions for reduced Stress Concentration Factor (SCF) T Joints of Circular Hollow Structures.* In: TIMA14 - Innovative technologies for joining advanced materials, 2014, pp. 44-49.

[20] DUGGAL , S. K. *Design of Steel Structures.* New Delhi: Tata McGraw Hill, 2009.

[21] GĂLETUŞE, S.; MITU, P.; STOIAN, Gh. *Construcţii metalice. Aeronave.* Bucureşti: Ed. Didactică şi pedagogică, 1978.

[22] GREEN, W.; SWANBOROUGH, G. *An illustrated anatomy of the world's fighters.* London: Salamander Book ltd, 1981.

[23] GROH, R. *A Brief History of Aircraft Structures.* Available on: aerospaceengineering.com, 2012.

[24] GROSU, I. *Calculul şi construcţia avionului.* Bucureşti : Ed. didactică şi pedagogică, 1965.

[25] GUTTMAN, R. *The Triplane Fighter Craze of 1917.* Available on: Historynet.com, 2011.

[26] HADDOW, G. W.; GROSZ, P. M. T*he German Giants - The German R-Planes 1914-1918* (3rd ed.). London: Putnam, 1988.

[27] HERTEL, H. *Leichtbau, Bauelemente, Bemessungen und Konstructionen von Flugzeugen und anderen Leichtbauwerken.* New York: Springer, 1980.

[28] HOFF, N. J. *A short history of the development of airplane*

structures. In: American Scientist, 1946, Vol. 34, No. 3.

[29] ILIESCU, P.; MITU, P.; STOIAN, G. *Manualul tinichigiului structurist de aviație*. București: Ed. Didactică și Pedagogică, 1974.

[30] JAKAB, P. L. *Wood to metal: The structural origins of the modern airplane*. In: Journal of Aircraft, 1999, Vol. 36, No. 6.

[31] JEGLEY, D.; VELICKI, A. *Status of Advanced Stitched Composite Structures*. In: AIAA Aero Sciences Meeting, 2011.

[32] KLEMIN, Al. *Aeronautical Engineering and Airplane Design*. Garner Moffat Co., New York, 1918.

[33] KOSIN, R. *The German Fighter since 1915*. London, Chicago: Putnam, 1988.

[34] KUROBANE, Y., et al. *Design guide for structural hollow section column connections*. CIDECT/ TUV Verlag, 2004.

[35] LOFTIN, L. K. *Quest for performance. Evolution of modern aircraft*. Washington, D.C.: NASA Scientific and Technical Information Branch, 1998.

[36] MARTIN, L. H.; PURKISS, J. A. *Structural Design of Steelwork*. Butterworth Heinemann, 2008.

[37] MEGSON, T. H. *An Introduction to Aircraft Structural Analysis*. Elsevier, 2010.

[38] NAZARI, A.; DURACK, J. *Application of the HSS Method to the Fatigue Assessment of HSS Shiploader Boom Connection*. In: the 5th Australasian Congress on Applied Mechanics, Brisbane, 2007, 10-12 Dec, pp. 211-217.

[39] NAZARI, A., et al. *HSS Design with parameters equations for fatigue assessment of tubular welded structure*. In: Australian Mining Tech.nology Conference, 26-27 Sept, 2006, pp. 301-309.

[40] NICA, Al., s.a. *Bazele fabricației navelor aerospațiale*. București: Ed. Tehnica, 1976.

[41] NICCOLI, R. *History of Flight*. Vercelli, Italy: White Star, 2002.

[42] NIȚĂ, M. M.; MORARU, Fl. V. ; PATRAULEA, R, N. *Avioane și rachete. Concepte de proiectare*. București: Editura Militară, 1985.

[43] NIU, M. C. Y. *Airframe Structural Design*. Hong Kong Conlimited Press, 1988.

[44] NIU, M. C. Y. *Aircraft Stress Analysis and Sizing*. Hong Kong Conlimited Press, 1997.

[45] PACKER, J. A.; HENDERSON, J. E. *Hollow Structural Section – Connections and trusses.* Allison: Ontario Canadian Institute of Steel Corporation, 1997.

[46] PARMLEY, R. *Illustrated Sourcebook of Mechanical Components.* London: McGraw Hill, 2000.

[47] PAUL, D., et al. *Evolution of US Military Aircraft Structures Technology.* In: Journal of Aircraft, 2002, Vol. 39, No. 1, pp. 18-29.

[48] PAZMANY, L. *Landing gear design for light aircraft.* SanDiego, US: Pazmany Aircraft Corporation, 1986.

[49] PETRE, A. *Proiectarea struţurilor de aero-astronave.* Bucureşti: Editura Academiei Române, 1999.

[50] POMILIO, O. *Airplane design and construction.* New York: McGraw Hill, 1919.

[51] PREOTU, O. *Construcţia şi calculul aeronavelor.* Bucureşti: Ed. Tehnică, 2001.

[52] PUNMIA, B. C.; ASHOK, K. J.; ARUN, K. J. *Comprehensive Design of Steel Structures.* New Delhi: Laxmi Publications Pvt Ltd, 1998.

[53] QUERIN, O. M., s.a. *Topology and Parametric Optimisation of a Lattice Composite Fuselage Structure.* Fort Mill: Altair University, 2013.

[54] RATHBURN, J. B. *Aeroplane construction and operation.* Chicago: Stanton and Van Vljiet, 1918.

[55] ROSKAM, J. *Airplane Design.* Ottawa: Roskam Aviation and Engineering Corporation, 1985.

[56] SALCA, H. *Contribuţii româneşti în aviaţie.* Braşov: Ed. Transilvania Express, Braşov, 2003.

[57] SCHUHMACHER G., Optimising Aircraft Structures, Concept to Reality/ Altair Engineering, 2006

[58] SENSMEIER, M. D.; SAMAREH, J. A. *A Study of Vehicle Structural Layouts in Post WWII Aircraft.* American Institute of Aeronautics and Astronautics, 2004.

[59] SHEYNIN, V. M.; KOZLOVSKIY, V. I. *Problems of designing passenger aircraft.* Moscova: Mashinostroyenie Press, 1972.

[60] SMITHSONIAN Institution's "Airspace Mag", www.si.edu

[61] TORENBEEK, E. *Synthesis of subsonic aircraft design*. Delft: Deflt University Press, 1976.

[62] VASILIEV, G. V. *Bazele calculului structurilor aeronautice cu pereţi subţiri*, Vol. 2. Bucureşti: Editura Academiei, 1988.

[63] VISSERING, H. *Zeppelin. The story of a great achievement*. Chicago: Wells &Co, 1922.

[64] Vivian E. Ch., A History of Aeronautics, Harcourt, New York, 1921

[65] Wardenier J., et al., Hollow Sections în Structural Applications, CIDECT, 2010

[66] WANHILL, R. J. *Milestone Case Histories in Aircraft Structural integrity*, NLR-TP-2002-521, National Aerospace Laboratory, 2002.

[67] WELLS, M. *A history of engineering and structural design*. Routledge, Oxfordshire, 2010.

[68] ZHAO, X. J., et al. *Design guide for circular and rectangular hollow section welded joints under fatigue loading*. CIDECT/ TUV Verlag, 2001.

[69] ZHANG, X.; LI, Y. *Damage Tolerance and Fail Safety of Welded Aircraft Wing Panels*. In: AIAA Journal, 2005, Vol. 43, No. 7, pp. 1613-1623.

[70] * * *, Aircraft Detail Design Manual, Aviation Publications, Appleton, US, 1977

[71] * * *, Aviation Maintenance Technician Handbok - Airframe, FAA-H-8083-31, Vol. I, FAA, 2012

[72] * * *, centuryofflight.net

[73] * * *, Dornier Post Sonderaurgabe, Dornier, Friedrichshafen,1984.

[74] * * *, Isogrid Design Handbook, McDonell Douglas Astronautics Company West, 1973.

[75] * * *, Air Cadet Publication ACP33, Vol. 4 Airframes, 2000, 967atc.co.uk

[76] * * *, Isogrid CNRC – IRAP, Aerospace Innovation Forum, Montreal, 2011.

[77] * * * , History of Aircraft Structures and Structural Design Considerations for Contemporary Aircraft, engr.sjsu.edu

[78] * * *, wikipedia.org
[79] * * *, wwiaviation.com

[F01] 1913-linner.ro
[F02] stonefoundries.com
[F03] jamco.co.jp
[F04] k-zeitung.de
[F05] reinforcedplastics.com
[F06] compositesworld.com
[F07] xyperion.com
[F08] lr.tudelft.nl
[F09] flightaware.com
[F10] flightglobal.com
[F11] ctie.monash.edu.au
[F12] Deutschesmuseum, Oberschleissheim
[F13] aviationgraphic.com
[F14]Popular Mechanics, 1912
[F15] amongruins.com
[F16] air-and-space.com
[F17] dartzkombat.com
[F18] Deutsches Museum, Munchen
[F19] natureandtech.com
[F20] ww2aircraft.net
[F21] Revista Flight, 1928-1932
[F22] airminded.net
[F23] yosikava.livejournal.com
[F24] The Aircraft Engineer, 1936
[F25] howthingsfly.si.edu
[F26] jneaircraft.com
[F27] iwm.org.uk
[F28] asisbiz.com
[F29] crimso.msk.ru
[F30] xtreemhost.com
[F31] combatreform.org

[F32] airforce.gov.au

[F33] bikeforums.net

[F34] framework.latimes.com

[F35] ccp.uair.arizona.edu

[F36] thethingswithwings.com

[F37] isogrid-sst.com

[F38] aerolinks.net

[F39] clipnuts.com

[F40] heritageconcorde.com

[F41] erthartwell.com

[F42] pprune.org

[F43] chiron.de

[F44] f-16.net

[F45] scanliners.com

[F46] bayourenaissanceman.ro

[F47] pilotfriend.com

[F48] stiintasitehnica.com

[F49] topology-opt.com

[F50] century-of-flight.net

[F51] flightglobal.com

[F52] waterandpower.org

[F53 76] fiddlersgreen.net

[F54] livescience.com

[F55] flitzerbiplane.com

[F56] mapairmuseum.org

[F57] cammfollowers.org

[F58] skynet.be

[F59] 967atc.co.uk/wordpress

[F60] flick.com

[F61] ruag.com

[F62] spitfirespares.com

[F63] hughtechnotes.com

[F64] jetsales.com

[F65] preventionaction.org

[F66] latimes.com

[F67] aeronauticausa.com

[F68] leasonslearned.faa.gov

[F69] militaryphotos.net

[F70] finemodelworks.com

[F71] biplaneforum.com

[F72] avia-it.com

Table of Contents

www.ingramcontent.com/pod-product-compliance
Lightning Source LLC
Chambersburg PA
CBHW080048240326
41599CB00052B/11